GW00702856

Michael Evenari

The Awakening Desert

The Autobiography of an Israeli Scientist

With a Foreword by Hermann Remmert

With 11 Figures

Springer-Verlag Berlin Heidelberg New York
London Paris Tokyo Hong Kong

Prof. Dr. MICHAEL EVENARI†
Hebrew University of Jerusalem, Israel

Title of the original German edition: Und die Wüste trage Frucht.
Ein Lebensbericht © by Bleicher Verlag 1987

ISBN 3-540-50794-9 Springer-Verlag Berlin Heidelberg New York
ISBN 0-387-50794-9 Springer-Verlag New York Berlin Heidelberg

© Springer-Verlag Berlin Heidelberg 1989
Printed in the United States of America

2131/3145-543210 − Printed on acid-free paper

This book is dedicated to my wife Liselotte,
a descendent of the ancient Jewish family Kalonymos.
Without her
this book could never have been written.

Foreword

In this book, first written and published in his mother tongue, German, one of the greatest living scientists, Michael Evenari, tells us the story of his life from the days in pre-war Germany to his participation in the struggle to create an independent state of Israel. At the same time it is the story of a scientist who never ceased to pursue his calling, however dire the conditions.

In the middle of the twentieth century, with the eradication of an ancient people, central Europe deprived itself of an irreplaceable part of its cultural heritage and potential. This is nowhere better illustrated than in the present book.

Michael Evenari has not only become a highly renowned scientist: He has served his country as soldier and, later, as ambassador in the cause of building up the Hebrew University at Jerusalem. By combining his vast knowledge of botany, soil science, ecology, history and archaeology he succeeded in making the desert green - and earned himself worldwide fame.

Despite our cooperation as editors of the journal "OECOLOGIA", and although I have visited his home in Jerusalem and his farm in Avdat, Michael Evenari's request that I, a German zoologist, and a much younger one at that, should write a short foreword to his autobiography, came as a complete surprise. Although it is difficult, who else should write it but a German?

And so I ask you to read this book, to read of the inhuman lengths to which an ideology, blindly pursued, can lead, and to marvel at what genius can achieve even under the most nightmarish of conditions. Once again it is brought home to us what we Germans have lost.

We thank you Michael Evenari for giving us this book and we thank Liesel Evenari for all her encouragement and stimulation without which this book would not have been possible.

Hermann Remmert, Marburg, Spring 1989.

Upon returning from a learning-teaching time in Poland I found Lieselotte Evenari's letter on my desk telling me about the death of Michael Evenari. Although not completely unexpected, this message hits met. The world — not only of biologists — has become poorer. Michael Evenari — a great humanist, a great fighter, a great botanist has died. We are called up to live and to learn from his life and his deeds.

Summer 1989.

Acknowledgement

My thanks are first of all due to Prof. Dr. Naphtali Tadmor and Dr. Leslie Shanan, who as a team with me carried out the research and reconstruction of our three runoff farms in the Negev. Thanks are also due to the various most loyal managers of our farms: Jacov Idan, Aryeh Rogel, Michael Sadeh and David Masig, to Udo Nessler and Otto Schenk, who helped to build up our runoff farm in Wadi Mashash, to the hundreds of volunteers from all over the world who worked at the farms, and to my secretary Tineke Kurtz who typed, retyped and corrected my manuscript. Last but not least, I thank the Rockefeller, Ford, Rothschild and Hillson foundations, the German organizations "Brot für die Welt", "Weltfriedensdienst", "Evangelic Church of Hessen-Nassau" and "Evangelic Churches of Switzerland", which financed our work at various stages, the Hebrew University of Jerusalem, the Ben-Gurion University of the Negev and the German "Forschungsgemeinschaft" (D.F.G.), which enabled us to carry out our desert research together with my friend Prof. Dr. O.L. Lange, the eminent eco-physiologist of the University of Würzburg and his team.

Michael Evenari

Contents

Prologue
How it Happened

In the early afternoon of the 5th of October 1965, with the heat of the summer still raging, suddenly, without warning, our experimental farm in Avdat in the Negev desert was completely surrounded by a wild torrent of flood water. We were completely taken by surprise as to the best of our knowledge, flooding never occurred so early in the fall. At the time a completely inexperienced student, who had just joined us, my wife Liselotte and I were alone on the farm. The other staff members had returned to their families in Dimona earlier in the day for Yom Kippur, the holiest day in the Jewish calendar.

That morning the sky had been clear and the weather pleasant, typical of an early autumn day in the Negev. Between five and six in the afternoon, the weather had changed drastically in a matter of minutes, a mass of dark, low-flying cloud obscured the entire sky from horizon to horizon and the heavenly sluice gates opened in a torrential downpour with rolling thunder and flashing lightning. Rushing out of the house, we were confronted by the first waves of the flood water streaming down the hills onto the terraces of the farm. The rain had not yet stopped and the water flowed endlessly from the surrounding hills. Clearly, if we did not act immediately, the farm would soon be totally flooded beyond its water holding-capacity. The water had to be channeled into the irrigation aqueducts.

Wearing high rubber boots and heavy raincoats, we ran down the hill to the fields. So great was the amount of rushing water, that we had to run from terrace to terrace to open and close the sluice gates channeling the water into the terraces. The loess soil had turned into soapy mud on which we slithered about. The force of the flood water was so great at one point that Liesel, who was trying to take photographs, was almost swept away. The student and I, busy at the other end of the farm, were unaware of the danger she was in.

The rain, thunder, and lightning continued unabated. The terraced farm became totally waterlogged looking like a lake. Fearing that the dams and walls would collapse and that the farm would be swept away, we tried to reinforce the endangered walls with heavy sandbags.

On returning to the house to change her film, Liesel found the telephone ringing. Aryeh, the manager of the farm, and his assistant Micky, had been trying to reach us wanting to wish us an easy Yom Kippur fast. When Liesel told them that the farm was under water, they found it hard to believe since in Dimona, only 55 kilometers north of Avdat, the sky was clear and not a single drop of rain had fallen. Both of them wanted to come to the farm to help us, but Liesel refused their offer as the road to Avdat, partly washed away by flood water, was dangerous.

When Liesel came back from the house we continued to try and save the farm. Night fell. We suddenly saw a light flashing beyond the water at the lower end of the farm. Aryeh and Micky had come. Since the road was under a meter of water, they left their jeep far away from the farm. Fortunately, owing to their good sense, they were spared the fate of many unsuspecting drivers, who every year are swept away by flash floods. Aryeh and Micky had walked to the farm and saw us with our flashlights. We warned them not to come any nearer because of the strong currents. Ignoring our warning, they stripped down to their underwear and swam across the "farm-lake".

Working together we saved the farm, and when the rain and flood waters subsided at about two in the morning we were able to leave the farm, which now had enough water for a year. Exhausted but happy, we went back to our house. We had succeeded in harnessing the flood water so that it could subsequently be used by the trees and crops of the farm.

After a considerable delay we sat down to the festive meal, which is customarily eaten at sundown before the fast. Afterwards, completely drained, I fell into bed, but was so exhausted that I could not fall asleep. In the curious state between sleep and consciousness, I had a "vision". I was again a child in the nursery of our house in Metz. I spoke to my childhood-self and asked "How can I simultaneously be a child, and an adult in Avdat?" Then, my childhood-self faded away. A voice inside me said "You will have to find an answer to that question some time in the future." Twenty years later, I embarked on my search.

Chapter 1
Youth

I was born Walter Schwarz on October 9th, 1904 in the city of Metz in Alsace-Lorraine, the two border provinces between France and Germany. Metz is an ancient town which was originally founded by the Romans who named it Divodorum Mediomatrici. Under Gallo-Roman, Frankish and Carolingian rule the city was an important center of trade and commerce. From the 10th to the 13th centuries it became the seat of several influential bishops. For the next 300 years it was an oligarchic republic until it was occupied by the French kings. During the Franco-Prussian war of 1870-1, after a long siege, Metz fell into German hands. The Germans united Alsace with a part of Lorraine hence, the name Alsace-Lorraine and subsequently made Metz the capital of Lorraine.

In my childhood the siege of Metz was still fresh in the memory of the townspeople. The elderly would recount stories of being so driven by hunger that they would hunt and eat rats that were still plentiful when I was a child.

Each epoch of the long history of Metz left its trace on the city. There were ruins of an amphitheater, the largest in the Roman province of Gallia, the remains of a 22-kilometer-long Roman aqueduct, churches, monasteries and a cathedral from the Middle Ages, palaces, large government buildings and imposing army barracks built during the 400 years of French rule. We children searched for artifacts in Gravelotte, Mars-La Tour and St. Privat, places where bloody battles had been fought in 1870 for the control of the city.

For me as a child, these remains of the past were alive. I saw myself as a Roman emperor visiting the city, a spectator in the amphitheater, the mayor of the Republic of Metz, a general fighting in the Franco-Prussian war. I know today that these childish games helped lay the foundation for my fascination for history which was to develop with the years.

My interest for history was also sharpened by the fact that the Jewish community of Metz had a history as long as that of the city itself. The first Jews had come to Metz with the Roman legions. In the Middle Ages the Jewish community of Metz was as important as those of Trier, Speyer, Mainz and Worms. In the 15th and 16th centuries

Metz was known in the Jewish world as the "birthplace of European rabbinic learning."

The importance for Man to be aware of his past is expressed eloquently by Laurens van der Post who wrote, "The human spirit cannot do without a past and unless it can feel itself putting its roots into its own past rather as a tree goes deeper into the earth for every inch it grows, it cannot move into the future." (1, p.281).

It is tragic that so many of today's youths have a tendency to negate the importance of their relationship to the history of their own nation, acting as if it means nothing to them. I became aware of this attitude when, during one of the many discussions on the Nazi regime in Germany, one of the German volunteers at our Avdat farm told me: "What the Nazis did does not concern me, I was not yet born at that time." My wife Liesel replied, "You are lucky. After 2000 years people still accuse me of having killed Jesus."

I was to discover this same attitude while visiting Russia whenever I mentioned the name Stalin. According to the people I spoke with he was a "non-person". These "forgetful" people should know better. Only being conscious of history can mankind realize the mistakes of the past and create a better future.

Since the occupation of Metz by Germany in 1870, German had been the official, and French the unofficial language. I was therefore bilingual from early youth. My French-speaking nurse and our maids imbued me with French, while my parents spoke to me in German. There was also a third "language" spoken in Metz. It was a combination of German and French. I still remember having heard the following sentence: "Maman, komm à la fenêtre. Jean ne croit pas, dass Du schielst." ("Mom come to the window, Jean does not believe that you are cross-eyed").

My bilinguality, together with the Latin and Greek I later learned in high school, helped me to master English, Spanish, Italian and Russian as an adult. Latin and Greek have always helped me immensely in understanding scientific terms and names of plants and animals. Knowing, for example, that "polyphyllus" means many-leaved, or that "pusillus" means tiny, helps in correlating the names with the actual plant or animal. My experience with Israeli students has only reinforced my belief that Latin and Greek is extremely helpful to a science student. Because most Israelis do not learn either language in school, it is often very difficult for them to remember scientific names and terms. Knowledge of English is indispensible for scientists today since it's the official language of most scientific congresses; the most important scientific publications are written in English. In addition, a scientist should have at least a reading knowledge of Russian. Many relevant papers and books published in Russian are unfortunately never read by scientists in the Western World because of the language barrier.

If scientists were as rational as they always claim to be, the knowledge of all these languages wouldn't be necessary. It would seem to be more logical to return to the system which was used during the Middle Ages through the Renaissance, using Latin as the official international language of scholars and scientists.

There exists at least one other valid reason for a scientist to learn several languages. When visiting Third World countries, scientists often have problems communicating information due to the language barrier. The more languages one knows, the less likely that this will be a problem.

During my childhood Metz was a place with divided national allegiances. Many French inhabitants had left the town after its occupation by German forces, not wishing to live under German rule. In their place, many Germans immigrated to Metz, enjoying the many opportunities that existed after the war. Because the city lay at the frontier between France and Germany it became an important military center stationing 20,000 - 30,000 soldiers permanently based there. For enterprising people, this meant new buildings as well as renovation of old ones. There was constant construction and business flourished.

My parents were part of this new wave of immigration. They opened a small shop which in a short time was to become the first large department store of Metz and Lorraine, called the "Kaufhaus Hermann Schwarz" (Schwarz's department store).

The French who stayed on in Metz also played a role in the economic boom, but the rift between them and the German immigrants grew deeper. The underlying reasons for this were not always political, and often happens in situations like this. The hostility was fuelled by seemingly trivial incidents. For instance, the Germans restored the beautiful old cathedral with the "Mutte", a well-known giant bell, and managed to annoy the French by ringing it after every victory by German Forces in the First World War. They also altered the features of the face of the statue of Daniel on the cathedral facade to those of William the Second, who was the emperor of Germany at that time. The French who were also good Catholics were understandably offended by this likeness of a German Protestant! After the armistice the first act of the French troops was to hang a sign round the statue's neck, bearing the words: "Sic transit gloria mundi" (Thus passes the glory of the world). During the war, the new law forbidding people to speak French in public further deepened the split between the two groups.

This division between the French and the German also affected the Jewish community. Everyone prayed in the same synagogue, but there were no intermarriages or friendships between these two groups. The close friendship between my sister Erna and a Jewish French woman, Alice Lazard, a notable exception to this unwritten rule, later played an important part in my life, since I met my wife Lieselotte through her. Only Hitler made us understand that we were neither French nor German Jews, but Jews.

Dr. Netter, Rabbi of Metz and Chief Rabbi of Lorraine, had a special place in the dynamic French-German allegiances: although a firm French patriot, he gave sermons in German during the war and was even awarded a medal by the German government. I can still see him at the pulpit, dressed like a Catholic priest in his long, black gown, decorated with the German medal delivering lukewarm "patriotic" sermons to celebrate German victories. After the armistice, the medal was never seen again, and his patriotic feeling toward France was so great that rather than send his only daughter to a German school, he chose a French convent, where the nuns succeeded in converting her to Catholicism. Dr. Netter never forgave his daughter for this and would not see her any more. She eventually became a nun in an order requiring their members to sleep in their own coffins.

My school, the "humanistische Gymnasium" (the humanist high school) was in a very old building that used to be a French monastery on the banks of the Moselle river. There, we learned Latin and Greek, but not modern languages and certainly not French. I was a good student thanks to our governess Anna Lehnhard, whom we called Mam'sell Anna. She gave lessons to me and my cousin, little Erna, who lived with us, (not to be confused with my elder sister Erna, who was ten years older than me. My parents had little time for us children. They both worked in the department store from 7 in the morning until 8 at night with only a brief break at home around midday, and the store was also open on Sunday mornings. We did not eat with the grown-ups except on Sundays so that we did not even see our parents at mealtimes. It was hardly surprising that my parents were strangers to me.

Mam'sell Anna took their place. Tall and thin, wearing an old-fashioned pince-nez through which she glared at us when we had done something wrong, she came from Mecklenburg in north Germany. Although she was very strict with us, we loved her as a mother. In 1917, when she left the family after 10 years, I was deeply hurt, and became so depressed and unhappy that I cried myself to sleep for many nights.

After supervising our homework, Mam'sell Anna would read to us, especially from books of her favorite author, Fritz Reuter, who wrote in "Plattdeutsch", a German dialect resembling English, which was spoken in her hometown. It is to Mam'sell Anna's credit that I was able to speak and write English without having taken a single lesson. The trauma I went through when Mam'sell Anna left, and the outbreak of war, led to a drastic decline in my scholastic progress. On the first day of the war, my father sent me to buy 10 and 20 Mark gold coins. Boys from our school were sent to collect gold coins as we were told that it was the duty of every patriotic German to contribute these to the government "Reichsbank". Most German shopkeepers willingly gave up their gold, but I will never forget the cynical smiles of the French ones when they refused our requests. Naturally, we did not know then that this collection marked the end of an era with the withdrawal of gold coins from circulation, the long period of economic stability came to an end.

Financial instability was accompanied by threats to our physical security. Since we were so close to the active front line in Metz, we felt its effect more intensely than in other parts of the German Reich. During the First World War, Metz was surrounded by a double ring of fortifications maintained by the German army, and the front line was only 20 kilometers away at Mont a Mousson. At night from the window in my room, I could see the flash of the cannon fire and hear the continuous rumbling of the guns at Verdun, 50 kilometers west of Metz. During the early days of the war, came my first taste of aerial warfare when French planes attacked the city with "Fliegerpfeile" ("pilot's arrows"), used on that occasion, and as far as I know, for the first and last time. One arrow pierced the visor of the doorman's cap in my parents' store. The children fought to get hold of these rare souvenirs! Later, French planes bombarded the city, and on one occasion hit an ammunition train. During these explosions we were confined to the cellar for hours, where we sat huddled together frightened to death. Once, a huge rat came and sat on a heap of coal and stared at us with its shiny black eyes, apparently as terrified as we were. Another time, a bomb hit an anti-aircraft machine gun stand near our house, and the soldiers manning it were torn to pieces; shreds of flesh and uniforms were left to hang on a rose bush there for days.

Metz was also cut off because all forms of communication were limited. We were not allowed to leave the city without a special permit from the military governor. Inter-city mail was censored, and the telephone service was suspended throughout the war in order to prevent espionage since all the troops passed through the town on their way to and from the front line. In 1914 we witnessed the German defeat in the Battle of the Marne at first hand. Regiment after regiment of tired German soldiers passed our house on that hot summer day, and we stood in front of the house and gave the soldiers water to drink out of large buckets. In Germany proper, this defeat was kept a secret.

Retired teachers were called back to our school to replace the younger staff serving in the army. Having incompetent teachers who could not keep discipline in the classroom, the general atmosphere of war, and Mam'sell Anna leaving us had a strong effect on me. I stopped doing homework and spent most of the time playing with a band of rowdy children games with bayonets, sabers and hand grenades in the empty trenches surrounded by barbed wire. As a result in the 1917/18 school-year, I went from the best to the worst student in the class. My parents finally lost patience with me at the end of the school year of 1918 when our entire class, myself included, went berserk. When the teacher left the classroom to attend a short meeting, we tore all the furniture to pieces The most fun of all was "bombarding" romantic couples rowing on the Moselle River under the school windows with inkpots (at that time we still used metal nibs and ink). The teacher could not believe his eyes when he came back to the classroom after a short meeting and immediately called the principal. The culprits were asked to step forward, but everyone remained silent except me, and to this day I do not understand what motivated me to confess. The principal looked at me disdainfully and said, "You unworthy scoundrel, leave the classroom immediately. Your parents will hear what you have done, and you will bear the consequences." Meekly, I took my things and left without mentioning that the whole class was involved, and my schoolmates let me go without uttering a word that the entire class was guilty. I was later told that they persisted in placing all the blame on me when the teacher asked if there were any accomplices. I became the scapegoat, and still wonder how the teachers could have been so naive as to believe that one boy could have done so much damage. Thinking back, the incident smells of anti-semitism, for I was the only Jewish student in the class. My father, the richest Jew in Metz, was asked to foot the damage, which, to his credit, he paid unquestioningly.

I had learnt my lesson. From then on I no longer trusted my schoolmates and never took part in any of their pranks. The school in a "consilium abeundi", "advised" me to leave. So, at the age of 13, I was sent by my parents to live with my sister Erna and her husband Gerson Stern in Berlin. This move proved to be the best thing that could have happened to me. Mainly owing to the positive influence of my brother-in-law, my short stay in Berlin was a turning point in my life, which ultimately shaped my future.

Gerson Stern was radiant, sensitive, gentle, and lovable with the soul of a poet. Since his father died young he had to leave school and go into business in order to help his widowed mother. Although he had little talent for business, he managed to make money that later allowed him to become a writer. However, made penniless by the inflation of the twenties, he was forced to return to the world of commerce. During the First World War he was not drafted since he had a chronic gall bladder infection. Instead, he served in Berlin in a government office responsible for allocating food supplies.

Erna met Gerson in Berlin in 1916. A slender, good-looking man, he was very attractive to women. Erna, who was twenty years younger than he was, fell in love with this forty-year-old confirmed bachelor, and they were very happy together. Gerson died in his eighties in Israel. Even in his last days, he looked young, and his eyes burnt with the spirit of a twenty-year-old.

In spite of his enforced business activities, Gerson never stopped writing. He wrote many lyric poems, some of which were published in Israel, and later in Germany after the Second World War, as well as two books. Gerson's book, *Weg ohne Ende* (The Never-Ending Road), published in 1934 at a time when many Jews deeply shocked by the depth of Hitler's anti-semitism, became a best-seller, having a deep impact on German Jewry. It recounted the story of the brutal expulsion of the Jews from Prague in 1745 by an edict of the Austrian empress Maria Theresa. By showing those who despaired that their case was not unique, and that the Jews had been driven out of places in which they had felt at home for many centuries many times before, it gave some measure of consolation to the German Jews.

Gerson, with his sensitivity and gift for offering guidance to young people, recognized the slumbering scientist in my inquisitive nature. Shortly after my arrival in Berlin, he presented me with

R.H. Francé's book, *The World of Plants: a Popular Botany* (2), which had a great impact, opening up a new world to me. I had not been taught Biology in the "humanistische gymnasium", so that any knowledge I had about plants came from our factotum, Johann from Pomerania, who had kept our small garden in Metz, but I learnt very little about plants from him. By reading Francé's book, I discovered that plants were living organisms with complex structures, metabolism, with the ability of adapting to a wide range of environments. In particular, the two chapters: "Invisible Technicians", and "The Plant as Mother", fascinated me.

The former described how plant structure follows similar principles to those used by architects in designing buildings. Thus, plants had "invented" techniques which the human mind only "reinvented" millions of years later. I was even more taken with the chapter presenting the plant as mother, which described seed formation, structure and germination. Doubting the accuracy of Francé's explanation of germination, I embarked on my first experiment. I took some beans from my sister's kitchen, planted them in a pot of earth, and then watered them regularly. A feeling of surprise and a deep sense of awe overcame me as the seeds germinated, and I saw their hypocotyls breaking out of the soil and the two cotyledons unfolding. At the age of 14 I was unable to analyze my feelings, but now I know that they derived from what J.C. Jung called the "collective subconscious". In many ancient civilizations and religions, the mythical germination came to symbolize death and resurrection - the rebirth of life in spring after its "death" in winter.

The ancient Egyptians buried their dead with special vessels containing germinating seeds. The early Christians also employed germination in a symbolic sense, as Paul's letter to the Corinthians shows: "When you put a seed into the ground, it does not grow into a plant until it dies first. And when the green shoot comes up out of the seed, it is very different from the seed you first planted. For all you put into the ground is a dry little seed of wheat, or whatever it is you are planting" (Corinthians I, 15: 36-37). According to the Talmud (Ketuvoth 111 b): "Rabbi Khiyya son of Joseph said, "The righteous will be resurrected in the time to come clad in his own cloth.... If a grain of wheat which is buried naked, germinates surrounded by many coats, the more so the just who is buried only in his shroud." There are many other examples in which germination is used to symbolize the mystery of nature and human existence. It is easy to understand why germination presented such an enigma to man before the age of scientific thought. Placed in soil and given water, a seed, which appears to be dead, showing no sign of life comes alive, then begins to unfold and develop according to a predetermined pattern.

After I had consumed Francé's book, my brother-in-law gave me a child's microscope, which looked like one of the first microscopes invented, without a substage mirror and condenser, objectives that screwed on, and which was focused manually by elevating or lowering the tube. This primitive microscope opened the door wider to the world that I first caught a glimpse of in Francé's book. I collected water from puddles, in which I observed Euglena flagellating, and Volvox rotating majestically. Overwhelmed by the beauty of these organisms, I became curious to learn more about them. Shortly afterwards, my brother-in-law gave me another book, "*The Secret of Plant Life*" by Adolf Wagner (3). The title page showed a picture of two bean plants, one climbing up a pole and the other, reaching out to it from the top of a broken pole. As I began writing this autobiography, I thought of that book, which I had not seen for over 50 years, whose title I had forgotten, only recalling the author's name, that it had been written before the First World War and, in particular the picture on the front page. I asked some friends in Germany to find the book, and they managed to locate it on the basis of my description of the climbing bean plants. I think that the picture imprinted itself so deeply in my memory because it made me aware that plants have sensors enabling them to orient themselves in space. I read the book in a few days, and then grew a bean plant to see if Wagner's theories were true - this served as my introduction to Plant Physiology.

Once I had absorbed Francé's and Wagner's books, and looked down a microscope, my decision to become a botanist was final. From then on, I thought of myself in good Latin style as "discipulus scientiae amabilis".

Looking back, I realize my attitude to nature, which has not changed since then, was crystallized by my first encounter with living organisms at the early age of 13. Over the 60 years of my scientific career, I never lost my admiration for and awe for the complexity, beauty and incredible organization of living things, and curiosity still drives me to try and learn more. Plants, as partners of the living world, were never mere objects of investigation, and I always had a kind of personal relationship with them.

My admiration for nature increased when Charlotte Henschel, a young, very good looking and intelligent friend of my sister with whom I was in love came with me to the observatory at Treptow (now in East Berlin). I saw Saturn surrounded by rings freely suspended in space, Jupiter with its moons, and the spiral nebulae in Orion. Looking through the telescope, I felt as if I was floating into infinite space. Enthralled with the beauty and harmony of the world, "I cried and prayed," according to the records in my diary.

I went back to school in Berlin. My sister came with me when I presented the report card from my previous school (with its unsatisfactory marks) to the principal of one of the local "humanistische Gymnasium". For many good reasons, I can still remember that he was called Przygode. After looking at my marks, he gave me a withering look and said: "Schwarz, you seem to be a really rotten fruit." Only after my sister explained to him the special circumstances in my former school did he admit me to the school. Six months later when I left his school my scholastic improvement earned his compliments.

In 1919, Gerson finished his war work for the German government. We went to live in a house in Feudingen (Sauerland) which had been in the hands of the Loewensteins, my mother's family for over 150 years. According to family tradition and records, one of our ancestors, the Marquis de Loewenstein, had been banker to the kings of France. My uncle Selig Loewenstein also told us that the so-called "pope from the ghetto" Pietra di Leone, came from our family. Later, Gerson decided to devote himself to his writing in a quiet environment, so we moved to Buchenau, a small village in Hessen between Marburg and Biedenkopf, and there we lived in the country home of uncle Juda Isenberg. This typically Hessen-style house was surrounded by a large garden accommodating a cowshed. I installed my microscope, a primitive pharmacist's balance and a set of laboratory test tubes in my room which was on the top floor. I investigated the microscopic world in the water from the brook behind the house, and carried out simple chemical analysis of plant ash. There, I acquired the foundations for my interest in ecology.

My uncle was a cattle dealer and I accompanied him on trips to other villages to buy cows. We would drive the animals back to his cowshed in Buchenau using sticks. During our walks through the expansive, dense beech forests, my uncle told me the German names of the plants we came across and I made my first herbarium collection.

I led an idyllic life in that small village. On Sundays, the women wore traditional peasant dress. There was no bakery where one could buy bread. Each family brought its bread and cakes to be baked for all the villagers in a large communal oven and the women carried the loaves home on large trays balanced on their heads. They did their washing in the river Lahn, and then dried it on the grassy banks. I took part in all these activities.

After several weeks of this bucolic, free life in Buchenau, I moved on to the nearby university town of Marburg, where I took private lessons to prepare for my return to school. I boarded with Dr. Kroh, a high-school teacher, who taught me Latin, Greek and Mathematics. Since Dr. Kroh also taught in school in the mornings, I had a great deal of free time to do as I pleased. I was now 15 years old, relatively independent, and tried to behave and dress like a typical small-town university student. I attended quite a few lectures on general philosophy, and philosophy of religion, and nature, visited the botanical garden and the department of botany in the center of town. In the botanical garden. The scientific Latin names and places of origin of every plant were displayed on plaques. This made me aware of the fact that plants grow in different geographical areas - the branch of science known as Plant Geography.

Once, I came into the Department's lecture hall shortly after Professor Peter Clausen, the Head of the Department, had given a lecture on plant taxonomy. Many charts showing morphology and the structure of various plants, drawn by the professor himself, were hung behind the dias. Slide projectors and other teaching aids did not exist at the time, and professors either had to buy charts or draw their own. Although obsolete now, these charts were often real works of art.

Standing in that lecture hall, I inhaled the smell of academia for the first time and asked myself whether I would ever attend such lectures as a "real" student and perhaps one day even lecture myself? It seemed to be a far-off dream, but the thought made me feel happy.

The following year, 1920, my parents moved from Metz to Frankfurt-on-Main. As loyal German patriots, they opted for Germany after the occupation of Metz by France. Consequently, the French government confiscated all their belongings and their store as compensation for war damage. My family should have been fully reimbursed for this financial loss by the German government, but they were only given a small sum.

Another reason why my parents left Metz was because Otto, my elder brother, who had volunteered to serve in the German army at the age of 17, had been killed in action on the French front. Had they remained in Metz, I would have eventually been drafted into the French army, an eventuality which my parents wished to avoid. Later, in Hitler's era, my parents' patriotism for Germany and the fact that my brother gave his life for the country were totally discounted. Like all Jews, they were persecuted regardless of their national allegiances.

After two and a half years at the "Kaiser Friedrich Gymnasium" in Frankfurt, I passed my matriculation examination. We were not taught Biology, Physics, or Chemistry, and only learnt the rudiments of Mathematics. Consequently, I lost all interest in my studies, and only did the absolute minimum of work in order to pass the examinations. My real interests lay outside the school framework.

One of them was our Bavarian maid Resi, a good-looking girl who was well-developed and rather attractive. She seduced me when I was 16 years old. We were alone in the flat, I was doing my home work and she was sitting opposite me leaning over the table. I could not help admiring her charm, so we kissed, and then went further.

However Resi was not my only interest. I joined the Jewish Zionist youth movement "Blau-Weiss" (Blue-White), an organization I had already gone on hikes with in the "Züge" (groups) in Berlin. During one of those hikes in the "Grünewald", we met some other Blau-Weiss members, who spoke Hebrew among themselves. Since my only knowledge of Hebrew was from Sunday school, which I found boring, this really impressed me. I could read Hebrew in prayer books in the synagogue, but like most of the assimilated Jews, I did not understand what I was reading. For me, Hebrew was a dead language reserved for religious services. This meeting made me aware that Hebrew could be a living language and could be used to talk about everyday matters unconnected with religion.

The Blau-Weiss group in Frankfurt was led by Erich Fromm, who later became a world-famous psychoanalyst. A lively bunch of youngsters, under Fromm's able direction, we learned Hebrew, studied the Bible and the Talmud, dabbled in Jewish mysticism, and read Buber's books on Chassidism, Karl Marx's *Das Kapital*, and the books of the anarchists Bakunin and Kropotkin. I will never forget May 1, 1922 when we marched through the conservative city of Darmstadt, one of us carrying the blue and white flag of the Zionist movement and another, the red flag of socialism, By turn, we sang Hebrew Zionist and German socialist songs. The people walking down the main road looked at us in bewilderment, probably thinking that we had just escaped from a lunatic asylum. Through Blau-Weiss, I also met for the first time Yiddish-speaking Jews from Eastern Europe, who appeared to be steeped in a kind of innate Judaism. They taught us Yiddish songs, which we loved to sing.

Anti-semitism finally motivated me to become a Zionist. Like many other Jews, I first confronted this phenomenon when youngsters in the street in Metz called me "sale Juif" (dirty Jew). I responded immediately, hitting them with my school bag. In the fight that followed, I was quite badly beaten up.

Another anti-semitic incident proved even more decisive in making me a Zionist. My brother Otto, four years older than I, was a member of the German youth movement, the "Wandervogel" (literally meaning the "hiking bird") in Metz. I still have his membership pin of a silver flying griffin against a blue background. The group met in our garden and, accompanied by guitar, we sang old German folk songs about "Landsknechte" (the mercenaries of the Middle Ages), and minstrel, and sentimental love songs. In 1916, the Wandervogel became anti-semitic, expelling all the Jewish members on the grounds that they were "non-German." How could we patriots, representing the best of German culture in Metz, suddenly not be Germans any more? I can still feel the shock experienced then. Fortunately this incident coincided with the first mention I heard of "Zionism", a word which was strictly taboo in our household.

During the Jewish holidays, Jewish soldiers fighting on the front came to our synagogue. We were allowed to invite three or four of them home to join us at the festive meal. On one occasion, the artillery soldier and doctor we invited were both Zionists. They explained the aims of Zionism to me, telling me about Herzl, the founder of the movement, and put forward the view that being a Jew means more than adhering to faith: the Jews as a religion had given the world monotheism, the ethical values in the Ten Commandments, and the Sabbath, the day of rest; as a nation we had a duty to return to the land of Israel, known as Palestine at the time, and, in particular to the city of Jerusalem, These ideas were like salve on an open sore. I now knew where I belonged. I was a Jew, and not a German. I no longer wanted to be part of a people that did not want me. My parents never accepted my new national identity, remaining German patriots and opposing Zionism even after Hitler's rise.

My confrontation with anti-semitism in my home-town was later followed by an incident far worse than anything I had experienced previously. As already mentioned, In Marburg I boarded with the family of Dr. Kroh, my private tutor. He was married to a young, attractive girl with whom I flirted a bit. Dr. Kroh taught German at a Marburg high school and was very interested in German philology and linguistics. He collaborated in a project designed to make records of all the dialects spoken in the villages in Hessen. Kroh was a good teacher, and I made good scholastic progress with him. Both husband and wife treated me well.

One evening after dinner we sat congenially in the large living room. Dr. Kroh smoked his long grandfather pipe that reached down to the floor. His wife was pregnant, and knitting a baby jacket. It was a cozy atmosphere. During the conversation, Dr. Kroh asked me if I knew what was written in the Talmud. When I told him that I did not, he informed me that according to the Talmud, Jews were allowed to exploit non-Jews and even kill them. He also claimed that Jews for the Passover holiday killed non-Jewish children and used their blood. Over the years, this old anti-semitic story of ritual murder had cost many Jews their lives, but I had never heard this abominable slander before and was absolutely dumbfounded. I told him that this argument could not be true since surely I would know about it as a Jew. "Ah," he said, "but have you ever read the Talmud?" when I said "No," he replied triumphantly: "That's it. You should know that only the rabbis know the real content of the Talmud, which they keep a secret from nice, decent youngsters like you. The German translation of the Talmud won't help you because all the passages which show the true face of Judaism are omitted. Read *"The Protocols of the elders of Zion"*, and then you will see that my views about the Jews are justified."

His wife, who continued with her knitting throughout this conversation, nodded her head in agreement. "Anyway," he added, "As I witnessed with my own eyes in the last war, all Jews are cowards and shirkers." This argument infuriated me since my brother Otto lost his life in that war fighting for Germany. When I mentioned that, he pooh-poohed it as an "exception."

I also got another "practical" lesson in anti-semitism in Marburg. In 1920, when Kapp organized an uprising against the newly-established democratic government, the government asked University students and high-school seniors to volunteer for a special batallion ("Zeitfreiwilliger") within the framework of the "Reichswehr." I joined, pretending to be 18 years old, because I wanted to prove to Dr. Kroh that Jews were not shirkers. One evening the "comrades" in my barrack started making anti-semitic remarks. Furious, I asked them to stop, and when they refused, I attacked them, and was badly beaten. I then challenged the leader of the group to a duel with heavy sabers. He refused with a cynical smile and said that as a Jew, I was not worthy of that honor. Looking back, I cannot understand why, but his reply hurt me more than the beating. It reflected how naive I was and how deeply ingrained the German mentality, especially the idea of honor, was in me.

The Blau-Weiss played a crucial role in my development since it helped me to break away from this very German mentality, and in its place, fostered a new identification and awareness with Jewish life and thought. Through them, I got to know Franz Rosenzweig and read his book, *Der Stern der Erlösung,* an unprecedented work, which Gershom Scholem described as the "central Jewish-religious creation of this century." I also met Martin Buber and Rabbi Nehemia Nobel and was inspired by Nobel's mystic sermons. Together with Ernst Simon, Erich Fromm and Nahum Glatzer, I took part in the Talmud lessons given by Nobel.

I also loved to hike with my Blau-Weiss group. We often marched through the night, in dark spruce, oak and beech forests of the Spessart mountains, the Bavarian Fichtelgebirge, and the Roen. On one occasion, we went down the river Main on a barge made out of large trees, and lived on it for three days. We visited the old Frankonian cities of Dinkelsbühl and Rothenburg ob der Tauber, that had remained as they had been in the Middle Ages, with their castles, city walls and rows of beautiful houses all intact. Rothenburg was of particular interest to us since it was the town where the famous Rabbi Meir had lived. We visited his house which had a Hebrew inscription on it in his memory. The talmudic academy which he set up there in the 13th century became famous throughout the Jewish world. In 1286 on his way to the Holy Land in order to die there, Rabbi Meir was arrested by the German emperor Rudolf and imprisoned in Ensisheim in Alsace. The emperor was prepared to release him in return for the large ransom payment of 20,000 Marks from the Jewish community. In his fear that such an act would create a precedent for future acts of terror the Rabbi forbade the Jewish community to pay the ransom money. Thus, he remained in prison, where seven years later, he eventually died. The emperor even refused to give his body to the Jews for burial. We all knew this sad story and said the Kaddish (the prayer for the dead) in front of his house in memory of this great scholar and brave man.

On one occasion, I went on a hike to the Bavarian Alps, south of Munich, with two of my friends, Abram Dombrovski and Jacob Bernstein. Far away on the horizon I saw the serrated chain of the mountains piercing through a layer of clouds. This view strangely moved me. I felt as if I had come across something familiar that I had never seen before, and instantly fell in love with mountains, a passion that has never left me. Whenever I find myself in high mountain ranges, I feel this urge to climb. Only mountaineers can understand the feeling of utter freedom, intense happiness and satisfaction upon reaching the summit after a long and difficult climb. When we got to the "Kampenwand", I climbed its steep face without any previous training. I was in my element, like a fish in water. My friends could not understand what had come over me. They stood at the foot below staring up at me, afraid I would fall. Later, on the same hike, I climbed the Totenkirchel in the "Wilder Kaiser" mountain range in Tirol. About half way up, I suddenly found myself in the midst of a fierce thunderstorm; the thunder and its multiple echoes reverberating from peak to peak terrified me. I found a tiny niche in which I took cover until the storm had passed. I was unable to reach the summit, and climbed back down to my friends, who had given up hope of ever seeing me again.

I later climbed many other mountains. In 1977, a heart infarct put an end to my mountaineering career, but did not extinguish my love of mountains and their flora.

Chapter 2
Student Years: The Young Botanist

In 1923, I began my studies at the Johann Wolfgang von Goethe University in Frankfurt. I majored in botany and zoology and took chemistry and physics as minor subjects. Naturally the interest and focus of my studies was in Botany. However, I told my parents that I was studying chemistry and wanted to become a chemist. Had they known my real intentions, they would have thought that I had gone mad. Botany was hardly a suitable occupation for a good Jewish boy. How could one possibly earn a living as a botanist when many doors in academia were closed to Jews because of anti-semitism? I only revealed the truth to them after I was awarded my Ph.D. in Botany. They were speechless, and even when I already had a position as Assistant at the German University of Prague, they still had doubts about whether I could make a future for myself in my chosen profession. In 1930, my mother came to see me during a German Botanical Society meeting, held in Erfurt. On that occasion she asked my former teacher, Professor Möbius, "Herr Geheimrat, does my son, a Jew, really stand a chance as a botanist in a German university?" Since I was working on the physiological anatomy of the stalks of heavy fruits, at that time, he quoted a well-known German proverb in reply: "Gnädige Frau, es sind die schlechtesen Früchte nicht, an denen die Wespen nagen" (My dear lady, wasps do not nibble bad fruit). In other words, she did not have to worry about my future since I was a "good" fruit.

When I began my studies, Möbius, the only Professor of Botany at the university, was also Head of the Department. In keeping with the general pattern at the time, the Department was small with only one Full Professor, one Lecturer ("Privatdozent" Friedrich Laibach), and one Assistant Johannes Hemleben). Möbius as Full Professor was the all-powerful "boss", who could do as he pleased within the financial limits. He was a real gentleman. The Assistant, Hemleben, was a good scientist and a very interesting man. During my fourth semester there, he left the university for his true calling as an Anthroposophist priest. Over 40 years later, when I met him again, he was Bishop of the Anthroposophic Christian community in Hamburg.

There were only eight students majoring in Botany so that we were free to call on the Professor and his Assistant any time. For those familiar with universities as they are today, it is difficult to grasp the privilege of being introduced to science in general and to Botany in particular with so much individual attention.

One requirement for the Major was a practicum based on a novel method developed by Möbius. He prepared handwritten dossiers, which we called the "Fahrplan" (railway timetable), in which he outlined in detail all the observations and experiments we had to carry out. Every student had a key to the Department and could work through the practicum at his own pace at any time of day or night. Möbius had daily sessions with each student during which he looked at our work and discussed mistakes. If a student showed a special interest in a particular topic, he would encourage him to conduct research in that area. In this way, I came to publish my first scientific paper before completing my Ph.D. studies (4).

During my fourth semester at the university, on a visit to the Botanical Garden, I noticed the unusual arrangement in the vascular bundles of the petioles in a plant originating from Central Asia. Möbius had no explanation to offer for this phenomenon, and said: "Herr Schwarz, why don't you find out for yourself?"

Möbius became a kind of "father mentor" to me, although I always addressed him with his full title, "Herr Geheimrat." In turn to him, we were "Herr Schwarz," "Fraülein Levison," and "Frau Weinreich." It was unthinkable at that time for professors to call students by their first names, as is now common practice in many American universities. Even among ourselves, we used the formal mode of address "Herr," and "Fraülein" - a foreign concept to most western students of today!

Möbius was a great humanist. His main botanical interests were Taxonomy, Developmental and Physiological Anatomy, the nature and origin of plant colors (5), and the history of botany. I take pride in the fact that my name (Schwarz) is mentioned twice in his book on the history of botany (6). Apart from Botany, he had a wide range of intellectual interests, centering around Goethe's views on Natural History and color, and classical writings. He published several papers in which he identified plants in the drawings on ancient Greek ceramic vases and vessels. Möbius also displayed great courage, as this quotation from a letter he wrote to the Mayor of Frankfurt in 1935, at the height of the Nazi era shows: ".....Presumably you have seen..... the giant anti-semitic placards on the facade of Gontard's house. As Frankfurters we should be ashamed...that Jew-beating is carried out in such a distasteful, disgusting manner. Frankfurt should be particularly grateful to its Jewish citizens for their contribution in many fields......may I remind you that Frankfurt could never have been a university town, had the Jews in particular not donated the funds for this purpose...." Someone leaked the letter to the "Stürmer", the notorious anti-semitic journal edited by Julius Streicher. Consequently, Möbius' successor as head of the Botany Department, barred him from the Department, which he himself had founded.

In addition to basic training in General Botany, Möbius also aroused my interest in Taxonomy, a skill which subsequently was very useful to me when I worked in Aaron Aaronsohn's herbarium in Palestine. He also helped lay the foundation for my lifelong interest in Physiological Plant Anatomy and the History of Botany. In his lectures he always mentioned the names of scientists, who had made new discoveries or proposed various theories, or scientific terms. I have always tried to follow his example, as I feel it is one of the duties of university lecturers to instill historical awareness in his students, and to emphasize that all scientists stand on "the shoulders of giants."

Fritz Overbeck, who followed Hemleben as Assistant, introduced me to Plant Physiology. In his course, I carried out my first scientific experiment on seed germination, which later became one of the focal points of my research. Under his guidance, I also conducted my first physiological investigation on Gurwich's so-called "mitogenetic rays," which were supposed to cause cell divisions (7). I also started another study on the mechanism underlying variegation (the phenomenon of white stripes or spots ln leaves (8).

Möbius was not a good lecturer, listening to him was rather like taking a sleeping pill, that we jokingly referred to as Möbiol. The aristocratic Zoology Professor, Otto zur Strassen, on the other hand, was an extrovert, and brilliant lecturer. As well as his students, the intellectual elite of Frankfurt, especially women, were attracted to this impeccably dressed, charming man, and attended his lectures. Like the rest of the audience, I was spellbound by his description of the evolution of the animal kingdom: in his view, the various stages in this process were all well-documented. Listening to him, I would ask myself whether everything can really be explained so easily. I still have my doubts about the Neo-Darwinist theory he presented, but my fascination with his lectures based on his elegant manner of speech, accompanied by his beautiful, skillfully drawn tables, stands out clearly in my mind.

Zur Strassen's seminars were also unique in the way we had to report on recently published papers. He commented on the content of our reports, as well as their manner of presentation. I can still hear him saying, "Herr Schwarz, please don't read your lecture, speak freely. If you read from a prepared text, you will bore the audience, even with the most interesting material. Please stand still when you lecture, look at your audience and don't walk around on the dais, which distracts the listener. Another thing: never draw on the blackboard and talk at the same time. First draw, and then turn to your audience to explain the drawing." Often, when listening to lectures of some of my contemporaries, I think to myself that it is a pity that they did not have a teacher of zur Strassen's caliber.

The botanical excursions were high points in my university career. Hiking through the Swiss Alps down to the North Italian lakes we encountered high mountain flora, alpine meadows, mixed forests of spruce, larch, oak and beech trees, chestnut forests, and finally Mediterranean flora in Italy. Trips like these, demonstrating how altitude as well as geographical latitude determine plant distribution, are "living textbooks" of Plant Geography.

In another excursion, we visited North Germany and the island of Helgoland, where I collected a herbarium of sea algae, which is now housed in the Hebrew University of Jerusalem. At one point of that trip, I fell fully dressed into the cold North Sea, and had to be fished out. My colleagues gave me so much grog to prevent me from catching a cold that I got drunk.

Some of these excursions were joint trips with the Department of Botany of the University of Marburg under the guidance of Professor Claussen, whose drawings I had admired when I was in Marburg with Dr. Kroh. His explanations were most remarkable: as well as naming the plants we came across, he also gave full descriptions of their distribution pattern, morphology, biology, flowering behavior, and special physiological features. I wrote down every word as I put the plants in my "Botanisiertrommel," a special metal box to protect plants from drying out, which was the botanist's trademark at that time, but has since become obsolete.

Apparently, Claussen was as pedantic in his personal life as in his drawings and explanations. Otto Stocker once related to me how on being asked to appraise the personal qualifications of one of his colleagues, Claussen was rather complimentary, but finished his statement with the following *non sequitur*: "But I hasten to add that his opinion regarding the phylogeny of the Cruciferae (mustard family) is absolutely wrong."

After four years at the university, I graduated with a Ph.D. at the age of 22, and, as was the practice at the time, underwent an oral examination after my thesis (9) had been accepted by the Faculty. In contrast to the current requirements, this was the only examination I had to take during the course of my studies at the university. I take pity on students nowadays, who are plagued by numerous examinations, which I suspect hinder rather than promote the learning process. This test which lasted an hour, included topics in Physics and Chemistry, but the emphasis was on Botany and Zoology. It was more of a friendly discussion than a strict "examination." Möbius asked the first question, "Herr candidate, what can you tell us about the evolutionary history of the plant kingdom?" Zur Strassen added: "Please include animals in your answer."

I dutifully recited what I had learnt from their lectures on evolution and the textbooks. I do not know how I plucked up the courage, but then I added that although I was convinced that it took place, it was difficult for me to accept all the conventional explanations of the mechanism of evolution. Visibly taken aback, both professors immediately asked me on what I based this opinion. Shortly before the examination, I had read Kropotkin's book on the evolutionary role of mutual help among animals (10), and Becher's book on plant galls (11). This led me to doubt that the struggle for survival was the only factor involved in natural selection. As an example, I cited the development of complex structures by plants for insects (galls), in which a great deal of energy was expended without any apparent benefit to the plant. I also asked whether there was really sufficient proof that the seemingly perfect adaptation of plants and animals to their environment was the result of a long series of accidental micromutations. We then had a three way discussion, Möbius and zur Strassen regarding me as an equal.

A few days after getting my Ph.D., I married Alice Ollendorff, a niece of Alfred Kerr, the well-known writer and critic. We had met 5 years ago when I was 17 and she was 28 while we were both doing volunteer work in the offices of the Frankfurt Zionist Organization. Our marriage was not very happy.

While I was at university, Germany went through one of the most critical periods in its history. Superinflation, reaching a peak in 1923 when the nominal value of the highest banknote was one billion marks (about 25 American cents), took place. At times, the mark would lose half of its buying power within the course of an hour. The inflation ended when one billion marks became one new mark unit from one day to the next. In attempts to evade the effect of continuous devaluation, people bought whatever merchandise they could the moment they received their salaries. During university vacations, in the mornings, I worked as a porter in the vegetable market. After I got my pay cheque, I immediately bought matches, which served as a type of stable currency. I would then exchange the matches for food, which was not easy, since food was very expensive and peasants preferred to hold on to their produce rather than sell it on the market. After work I went out of town to the local villages, where I would try to trade other goods for potatoes or whatever food was available. Such exchanges were strictly forbidden, but driven by necessity, thousands of people went through similar rigmaroles. On our return to the railroad station, we all tried to evade the police check points. If caught, the goods were confiscated. I can still picture those huge crowds milling around the entrance to the railroad station, desperately trying to get past the police control.

One had to experience this terrible inflation in order to fully appreciate the impact it had on German society. Undoubtedly, it contributed to Hitler's rise. From one day to the next pensioners, even those with saving accounts, became paupers. People no longer knew how to cover basic living expenses. Confusion reigned everywhere and everyone was afraid of the future. They lost confidence in the government and other official organizations. Since the source of the galloping superinflation was unclear, people became suspicious. According to the rumors spread by Hitler, Jewish capitalists were responsible. Thus, volatile emotions of uncertainty, fear and suspicion nurtured the poisonous plant of Hitlerism.

After completing my doctorate, I worked as Assistant to Möbius for a year, until Ernst Pringsheim offered me a similar position in the Department of Plant Physiology at the German University of Prague. I worked there from 1927 to 1930.

At the time, Prague had two universities: a Czech and a German one. They both had Departments of Botany. The German university was at the top of a hill, and the Czech one at the bottom. Although very close to each other, there was very little contact between these two institutions. I only visited the Czech department once when I needed some information from Nemec, the well-known Czech Plant Physiologist. This lack of contact reflected the relations between these two peoples. Although these two nations lived together there was very little interaction or cooperation between them, foreshadowing Hitler's takeover of Czechoslavakia, which occurred ten years later. I was familiar with such states of affairs from my childhood days in Metz. I tried to enjoy the best of both cultures, since they each had much to offer. Under the Liberal-Democratic President Thomas Masaryk, there were many outstanding writers, poets, dramatists and musicians in Prague. The Czechs included people such as Karel Capek, Jaroslav Hacek ("the good soldier Schweik"), and the poets Victor Dyk and Vladimir Vancume, and the Germans, Franz Kafka, Franz Werfel, Stefan Zweig, and Ernst Brod, who were all Jewish, but wrote in German; they created a culture that was not typically German, but uniquely "Praguean" and which has since disappeared.

The Slavic culture encountered in Prague induced me to learn Russian. I decided that an educated man should know at least one Slavic language. I had already learned enough Czech to make my way around the city. I decided to learn some Russian because, in contrast to Czech, millions of people speak the language and it has rich literature. In addition, the many important botanical books and papers, written in Russian, would otherwise have remained inaccessible to me. Certainly, my knowledge of Russian opened a wider cultural horizon to me. I took lessons from Tanya, who was a niece of a Russian Professor of Philosophy, one of the many socialists and intellectuals who had fled to Prague from Bolshevik Russia. I fell in love with Tanya, which probably explained why I made such rapid progress in learning the language. Tanya's methods were quite original, rather than dwell on boring grammatical aspects, she taught through conversation and poetry. We read the poems of Lermontov and Pushkin, among others, some of which I can still recite by heart. This provided an excellent introduction to both use and the spirit of the language.

My stay in Prague also contributed significantly to my scientific development. I assisted Professor Pringsheim with his research, but also had ample time for my own scientific work, which I carried out under his direction. I focused on two problems: variegation, and the physiological anatomy of the fruit stalks of heavy fruit. I made the observations that the mechanical strength of fruit stalks such as those in pears and apples, were disproportionate to the weight of the fruits which would remain suspended on trees even with much less mechanical tissue in their stalks. Not the fruit stalks, but the connection between them and the twigs is the mechanically weak point causing the fruits to fall from the tree. From this I learnt an important scientific lesson, although various structures seem to fulfill certain functions, they are merely functionally neutral by-products of physiological processes not directly related to the observed structures (12, 13).

My stay in Prague also broadened my botanical horizons since I encountered many new topics there. I learnt a great deal from my two Departmental colleagues, Victor Czurda and Felix Mainx, who worked on the physiology and genetics of microscopic algae. I attended the lectures given by the famous algologist Adolf Pascher, Fritz Knoll, who was well-known for his work on the biology and ecology of flowers, and the plant sociologist Karl Rudolph.

In 1920 Heinz Oppenheimer, a botanist from Palestine came to spend several months in the Department. He had received his doctorate for a thesis on substances inhibiting seed germination in fleshy fruits such as tomatoes from Hans Molisch in Vienna University. He then emigrated to Palestine, where he began working in Aaron Aaronsohn's herbarium in Zikhron Ya'akov. His arrival in Prague was to change the course of my life.

Aaronsohn had a remarkably strong personality as well as being an autodidactic genius. Born in Rumania in 1876, he emigrated to Palestine with his parents at the age of six. Together with several other Rumanian Jews, his family founded Zikhron Ya'akov, one of the first Jewish agricultural settlements in Palestine. While helping his parents turn the stony, barren soil into agriculturally useful land, he developed an interest in the native plants of the surroundings. His scientific curiosity was further aroused when his uncle gave him a microscope, which he taught himself to use. In the small village school, his outstanding intelligence and abilities were soon recognized. His talents came to the attention of Baron Edmond de Rothchild, benefactor of the newly established Jewish agricultural settlements in Palestine at that time. The Baron decided that Aaronsohn's special abilities should be nurtured and therefore sent him to an agricultural school in Grignon, France for two years. He was not allowed to complete his matriculation examinations, however, because the Baron did not want young people to be tempted to remain in France. His education in France not only made him into an efficient agriculturalist, but also stimulated his interest in the natural sciences.

When he returned to Zikhron in 1896, he began organizing a herbarium of the plants that he had collected in Palestine, Lebanon, Syria, and Jordan. Only using the books on the flora of the Middle East written by Post, and Boissier, he managed to determine the scientific names of these plants, and with uncanny instinct knew immediately when he had come across a plant which was either new to the area, or previously unknown to science. Since Zikhron was not close to any scientific center, he could not compare his findings with similar species in other herbaria. However, he painstakingly labelled what he thought to be a new species. He also noted where he had found the plant, and described its habitat, the type of soil, and geological formation. Such precision was quite unusual in his time. He eventually became a self-taught botanist, and a bona fide geologist. The first geological map of Palestine was largely based on his observations.

In 1906 in the Galilee area, he made his most important discovery, which earned him international fame, of the "mother of the cultivated wheat", a wild wheat species from which most cultivated varieties are derived. He first reported this finding to Otto Warburg, a German-Jewish botanist, who later became head of the World Zionist Organization and founded the Botany Department of the Hebrew University, and Georg Schweinfurth, the well-known German explorer of Africa. Recognizing the theoretical and practical importance of his discovery, they reported it to David Fairchild, Head of the Division for Plant Introduction in the United State Ministry of Agriculture, and E.W. Hilgard, Head of the Department of Agriculture of the University of California at Berkeley.

His finding was so important that he was invited to lecture at several universities in the United States, where he also got to know many influential Jewish philanthropists. Recognizing his genius, they donated funds for the establishment of an agricultural research station, which was set up in Atlith, close to his home in Zikhron in 1910. Consequently, Aaronsohn was not only the first Jewish botanist and geologist in Palestine, but also the first scientific agriculturalist. Unfortunately, his station only operated for five years because he got drawn into politics during the First World War and the station was subsequently destroyed by the Turks in retribution for these activities. Only a few of the records and documents, describing his brilliant, original ideas and his preliminary research on ways to improve the primitive agriculture of the Middle East, have survived. At that time, Palestine was part of the Ottoman Empire. Aaronsohn realized that the oppressive rule of the Turks threatened the existence of the Jewish settlements. He feared that following the merciless massacre of millions of Armenians in 1915 by the Turks the next victims of Turkish brutality would be the Jews of Palestine. He felt certain that the Jews would never be able to realize the Zionist dream under Turkish rule, and decided to do all he could to help the British based in Egypt to overthrow the Turks in Palestine. He founded an organization, which he called NILI, an acronym for the Hebrew phrase in 1. Samuel 15:29: "The glory of Israel will not lie", with the purpose of providing the British headquarters in Cairo with information about the Turkish army in Palestine regarding their plans to attack the Sinai and the Suez Canal.

The espionage carried out by NILI was absolutely crucial for the British success, as the following statement made by Lord Allenby, General of the British Army in Palestine, shows: "The victory over the Turks was due to a large degree to the uncomparable work of Aaron Aaronsohn" (quoted from Florula Transjordanica, 14). Aaronsohn was honored for these activities by being invited to participate in the Peace Conference at Versailles. On that occasion, he insisted on the importance of a Jewish people's state in Palestine. Shortly afterwards on his way to England, the British military plane in which he was traveling vanished without trace. His disappearance remains a mystery to this day. As a first class intelligence officer, did he know too much?

Aaronsohn's greatness is succinctly expressed in a letter written in 1920 to his brother Alexander by W.C. Bullit, one of the most influential counsellors to President Wilson at the Peace Conference of Versailles and a man who was not given to exaggeration: "He was, I believe, the greatest man I have known. He seemed a sort of giant of an elder day-like Prometheus...Aaron to me was not merely the flaming embodiment of the determination of the Jewish race to have a home and to be again a nation, but rather a captain in the foremost company of that small army of humanity which marches even against ignorance, superstition and hatred. If he had lived I believe that he would have been a world leader through the years of travail which are now upon mankind".

After Aaronsohn's death, Heinz Oppenheimer was asked by his family to make an inventory of his plant collection, and to edit the list of the plants he had collected in "Transjordan" and the diaries, which he had written during his trips there. The results of this work were published in the book, "Florula Transjordanica" (14).

While in Prague, Oppenheimer stayed in the same boarding house as us, and we became friends. He asked me whether I would be interested in coming to Palestine to help to edit the list of plants in Aaronsohn's herbarium and his unpublished diaries (15). As I was already a convinced Zionist, his offer attracted me, and I promised to consider it seriously.

In the meantime, I was offered a post as Assistant in the Botany Department at the Technical University of Darmstadt, which I accepted as it gave me opportunity of becoming a Lecturer. In 1931, we moved to Darmstadt. I felt sorry about leaving Prague, "the golden mother," since the three years I spent there had been truly "golden". I enjoyed myself to the full, shutting out all thoughts of my depressing marriage.

Friedrich Oehlkers was Head of the Department when I first joined the university at Darmstadt. He was an interesting, intelligent, sensitive and nervous man with a deep interest in Philosophy. He was physically handicapped having lost an arm during the First World War. On one occasion, when I disturbed him unintentionally, he threw a chair at me, a quite unusual behavior for a professor. A snobbish person, he treated me and Kretschmar, his two Assistants, with disdain. In contrast to Möbius, he never invited us to his home, or showed the slightest interest in our private lives or scientific work. I gained very little from working with Oehlkers.

At the end of 1931, Oehlkers left and Bruno Huber was named Head of the Department. Oehlkers' aloofness was replaced by the warm affection of Huber. He encouraged me to write up the thesis for my lectureship, and I gave my trial lecture in 1933, although he was aware that I was Jewish and that the new Nazi regime was fanatically anti-semitic. He later got into trouble with the regime on account of what he had done for me.

I was responsible for organizing the laboratory exercises and floristic excursions for Pharmacology students and those training to be Biology teachers. In my first meeting with my students, I noticed that they all wore National Socialist Party badges. In order to avoid misunderstanding, I immediately made my position as a Jew and Zionist clear to them. Visibly taken aback, the students received my unexpected declaration in utter silence.

The relationship that subsequently developed between us showed that I had made the right decision. I was accepted as a Zionist, who wished to emigrate to Palestine, and we talked freely about Judaism, Zionism, and Nazism during the many nights we spent together at youth hostels while on botanical excursions. Sometimes, the discussions, particularly concerning their attraction to Nazism, which I considered to be the incarnation of evil, were very heated. Since I had read Hitler's book *Mein Kampf* (16), I considered in 1931 Hitler's becoming leader of Germany to be a strong possibility. However, most people who were aware of Hitler's rise did not read this "Bible of Nazism," and had they done so, would have considered it to be gibberish. On the other hand, I took his plans, clearly described in the book, very seriously. I felt sure that he would be spurred into action by his irrational hatred of Jews and Judaism, although in my wildest imagination I did not foresee the Holocaust.

In the early years of the Nazi regime, most German Jews were unwilling to recognize the dangers of anti-semitism, and did not face the situation until it was too late. I could not even convince my own family. After several discussions with my students, I tried to explain to my father what I envisaged would happen to the Jews when Hitler came to power. My father answered: "First of all, food is never eaten straight after it has been cooked. The Nazis make a lot of noise, but will be slow to act. Besides, we are Germans, and my son sacrificed his life in battle for the German Reich. As a good German patriot, I opted to leave Metz for Germany, and gave much of my wealth. Nothing will happen to *us*." He could not have been more mistaken.

In my discussions with the students, I tried to understand why they were so attracted to Nazism. As scientists, they were rational and intelligent, and they did not belong to the Nazi street gangs.

Gradually, I began to understand their motivation. After the bitter experience of super inflation, poverty, gave rise to uncertainty and loss of confidence in the state and government of the early twenties, a second great economic crisis befell Germany in 1929 as the consequence of the crash of the New York stock exchange. Bankruptcies in business and industry increased daily and gave rise to a terrible depression, culminating in the rise of the registered number of unemployed to 6 million in 1932. Wages fell and a large part of the population lived in misery and poverty. Everybody was affected, including my students. For the Nazis the depression was the indispensible condition for their rise to power. They exploited the situation, convincing the people that only they could solve the immediate problems of unemployment and poverty, which were deliberately brought about by World Jewry in order to dominate the world. My students served as a good example of the success of the Nazi propaganda machine.

Once, during a discussion on the distressing economic situation in the country, one of the students said "Herr Doctor, do you know why our party calls itself the "German National Socialist Workers Party" (N.S.D.A.P.)? We are the true socialists, not the so-called "international socialists". I don't know if you have read our socialist bible, *Breaking the Bonds of Interest Slavery* by Gottfried Feder (17). In that book you will see how our movement can make Germany a paradise for socialist workers." I had read Feder's pamphlet, published in 1919, and also remembered that in *Mein Kampf*, Hitler mentioned having heard Feder speak in 1919: "Immediately after hearing Feder's first lecture, the idea that I had finally found one of the most important principles for the establishment of a new party flashed through my mind.... When listening to Gottfried Feder's first lecture entitled, "Breaking the Bonds of Interest" I immediately knew that the issue in question was of fundamental importance to the future of the German people" (16). These lines demonstrate the extent Hitler's uncanny demagogic instinct perceived Feder's "socialism" as an unfailing attraction by the suffering masses of Hitler's uncanny demagogic instinct and their perception of Feder's "socialism". Clearly the name Hitler gave his party was a combination of the title "The National worker's party", which was already in existence at that time and a plagiarism of Feder's ideology. What exactly was Feder's socialist theory that attracted so many people? He postulated that there are two kinds of capital: the "good", "productive" industrial capital, which creates work, and the "bad", "unproductive" capital in the hands of rich people, most of whom were Jews. Wealthy people use this "unproductive capital" to lend money in return for interest, thereby ultimately enslaving the masses. The new Nazi state would abolish this "bond of interest slavery," and in doing so, enable the working man to borrow money without interest for buying houses, cars and luxuries. In the new state, both direct and indirect taxes would be abolished. Feder explained that the state income in return for services and the "good" industrial capital would provide sufficient funds to support this worker's paradise.

All efforts to explain to my students that the whole idea was an impossible bluff were in vain. I tried to show how it is impossible to draw the line between "good" industrial and "bad" private capital, and as long as the monetary system existed, there would always be interest. I also pointed out that if they were to apply a "scientific" approach to economic matters, they would soon realize how naive Feder's ideas were. But, it was to no avail, they clung firmly to Feder's brand of "socialism." I am absolutely certain that Feder's socialism was a major, and possibly the main attraction of Nazism for them.

Nowadays, when I discuss Nazism with German youth and even older people, mentioning the crucial role that Feder's book played in the success of the Nazi party program, they look at me in astonishment. They had not even heard of Feder. Although Hitler used socialism and Feder's ideas to lure people into the movement, he never really intended to incorporate these concepts into life. With the brutal murder of Gregor Strasser, a "socialist" who was his second-in-command, the socialist program vanished into thin air, and was forgotten by the German people. I always sensed that Hitler was only using socialism for political ends, and that he would drop it, together with Feder, like a hot potato, after his rise to power. To this argument, my students answered, "Oh no. Feder's ideas are important and Gregor Strasser would never allow the program to be phased out." They must have realized how wrong they had been when my prediction came true, and was reinforced by the decision to restore absolute power to the employers, taken by Dr. Ley, leader of the so-called Nazi "Labour Front", later that year.

"Socialism" was not the only bait used to lure my students to support Nazism. By joining the party, they not only identified themselves as "socialists", but also as "nationalists," "Germans," and "workers." Whenever the name of the party was mentioned, their eyes would light up with fiery enthusiasm and pride. This made me understand that even the name of the party was a demagogue's cunning of genius.

Being a member of the "master race" was another major attraction to Nazi ideology. In *Mein Kampf*, Hitler wrote: "The human culture we observe today, art, science and technology are almost exclusively the results of creative effort of Aryans. This fact alone provides sufficient bases to reach the conclusion that the Aryans established higher humanity, or the prototype of what is implied by the word "man". "In this world, human culture and civilization are inseparably linked with the Aryans. Thus death or decline of the Aryans would mean the dark veil of a lack of culture would once again descend upon the earth." "We all sense that in the distant future man will be confronted with problems that can only be solved by a master race...."

Thinking back, it never ceases to amaze me how I could discuss the explosive topic of the "Aryan race" with the students without coming to blows with them. Their answer to these arguments went something like this: "Herr Doctor, being a Jew, you can't possibly understand these things." I also spoke to them about Hitler's fanatical, pathological hatred of Jews, to which he devoted 33 pages in *Mein Kampf*. My students, who were not "rabid" anti-semites" like Hitler, but only "racial anti-semites" in that they considered the Jews a race to be foreign to Germany, claimed to like me as a person. In their Darmstadt dialect, they would say, "The Jews are not German and must leave Germany." For this reason, they considered it appropriate that I was a Zionist and wanted to emigrate to Palestine. On one point, we all agreed: the moment Hitler came to power, he would expel all Jews from Germany. At the time, none of us imagined that Hitler's "Final Solution" would mean the murder of six million Jews.

Since I never doubted even for a moment that Hitler would soon come to power, understanding Nazism and the general political and social atmosphere prevailing in Germany was very important. Consequently, I took these discussions with my students, and what I read in *Mein Kampf* very seriously. All traces of doubt about Hitler's imminent rise to power, swept away when my mother and I attended one of his mass rallies at the "Messehalle" in Frankfurt, a large hall that could seat thousands of people. Although the "SA" and "SS" guards at the entrance were under orders to beat up any Jews who tried to get in, we managed to enter the hall. I wanted to see and hear him speaking in order to better understand his magic hold over the masses. By chance, we got seats at the front opposite the "Führer."

Hitler was an unbelievable orator. It is difficult for those who never actually heard him speak to understand the spell he cast over his audience. As he came onto the stage, this rather ordinary looking man, underwent a dramatic change of personality. When thousands of the people rose in unison, shouting "Heil Hitler", and "Sieg Heil", he stood straight as a ram rod in the pose of a Roman emperor looking round at the screaming masses. Only then did I become aware of his blazing eyes, the outstanding feature in his face. Then he raised his right arm and the masses fell silent. He started to speak in a low, melodious voice, enumerating the terrible injustice to the German people of the Treaty of Versailles, the occupation of the Rhineland by the Allies, inflation, and the "so-called" democracy of the Weimar Republic. Raising his voice abruptly he shrieked: "Who is to blame for all this?" In his answer, he unequivocally blamed the Jews and "Jewish international capital" for the evils that had befallen Germany.

Then came the fireworks. A huge wave of shouting, screaming, and a flood of hatred poured over the audience. This demonic hate storm was not based on his words alone, his eyes also radiated hate waves, and with accompanying violent hand movements, he accused the invisible foes. Looking around me I saw men staring at Hitler with rapt, trancelike expressions. The womens' faces bore an expression of revelation, looking at their leader with religious adulation as if he were the Messiah with tears flowing down their faces. Some of them became hysterical while others looked as if they were experiencing orgasms. The atmosphere was loaded with emotion. I must confess that even I was moved by his oratory, although I knew that every word was a falsification of history, or a promise that he could not possibly keep. His booming voice and fiery eyes seemed to dull the reasoning faculties of his audience. Towards the end of his speech, he suddenly lowered his voice to a more or less normal tone. Raising his head as if speaking to God, he said: "And I myself and the party give you the solemn promise to change all this, to drive out the devils, and to lead our beloved homeland to the glory it deserves." For a few seconds, the audience remained silent. Then the dam of the accumulated emotions burst and a frenzied pandemonium of shouting, trampling, and screaming broke out. Possibly we were the only Jews to have ever attended such a rally, I am convinced that had more Jews, or non-German gentiles, heard such speeches, they would have realized that this demagogic demon could only be halted by force. If the English and French prime ministers Chamberlain and Daladier had seen Hitler's face at that mass rally, they would have never been duped by his guise of respectability in Munich, and never would have said, "Here is a reasonable man."

Both German Jews and the rest of the world were blind to Hitler's real nature, and refused to believe what was happening. My father for example went so far as to say: "We Jews should vote for Hitler. The faster he comes to power the sooner this whole spook will be over." The political leaders in England and America were blind to the reality. The first unabridged English translation of *Mein Kampf* only appeared in 1939, fourteen years after the German original. Ironically, the publishers note in the English edition states: "...a book which is likely to remain the most important political track of our time" (18).

Even with the publication of *Mein Kampf* in English, it seems that the Anglo-American political establishment did not bother to read it. Had they done so, Hitler's political aims would have become clear to them from the first page of *Mein Kampf*. "Only when the borders of the Reich include all Germans, and when it is no longer possible to assure him of daily bread, there arises out of national distress the *moral right* to acquire foreign soil and territory. The sword then becomes the plow, and the tears of war will produce the daily bread for generations to come.... A state, which in this age of racial pollution cherishes the best racial elements, must ultimately become the master of all the world."

Mein Kampf is also full of historical lies. Had thinking, rational people read his work, the falsification of history and Hilter's true intentions would soon have been apparent to them. I will only quote a few examples of the lies directed against Jews and Judaism. "The Jews were always a people and never a religion". "The Jew cannot possess a religious institution since he lacks any form of idealism and does not believe in the world-to-come." "The Jews never had a state with defined territorial boundaries, and consequently, never had a culture of their own...." Once Hitler came to power, he continued using distortions of the truth in order to manipulate the people, and primarily, to "justify" his acts of aggression, such as the occupation of the Rhineland and Czechoslovakia, his attack on Poland, and the pact with his Bolshevik archenemy, Russia, which he later violated.

I had kept up my contact with Oppenheimer in Palestine and when it was clear to me that the situation in Germany was getting worse, I decided to sign a contract with the Aaronsohn family in Palestine, and to go there in October 1933.

The period of almost two years, which I spent with Huber in Darmstadt, was quite beneficial professionally. I had observed that variegated leaves of a species of Coleus rooted easily without forming buds, the white sections becoming green. Based on these findings, I formulated a theory to explain this type of variegation. By far the most interesting observation I made in these isolated rooted leaves, which I was able to keep alive for months, was that the petioles formed a cambium (tissue from which new cells are formed), exhibiting the anatomical features and functions of stems. I also discovered that in many other species such as the Oleander, leaves can also root in a similar manner. My studies relating to these phenomena and their underlying physiology were the subject of my "habilitationschrift" (19), the paper I had to submit in order to become a lecturer. With Huber's encouragement, I presented my probationary lecture in March 1933. I then went on to investigate the development of plastids in young cells of shoot meristems. I think that I was the first person to observe mitochondria and proplastids *in vivo*, and to discover that they are completely independent both physiologically and developmentally. At the time this was important work, since some scientists did not believe that mitochondria existed, while others thought that plastids were formed from mitochondria (20).

Under Huber's influence I became interested in Ecophysiology. He was the first to use a fast-weighing torsion balance for measuring transpiration in field experiments, and he also developed a cumbersome instrument for the field measurements of photosynthesis. He recommended Maximov's book (21) "*A Study of the Physiological Basis of Drought Resistance*", a topic which I still work on today. In 1932, when I told Huber that I intended to emigrate to Palestine, he suggested that I take equipment for carrying out ecophysiological work in the Palestinian desert with me, and gave me two "desert bibles" on the physiological anatomy and water balance of desert plants, one by Volkens (22) and the other by Stocker (23). Otto Stocker was the first ecophysiologist to work in an Old World desert region. I found his work very inspiring, and decided to try and follow their footsteps when I got to Palestine.

Later, I became personal friends with both Huber and Stocker, and they came to visit us at our desert research station at Avdat. Stocker came nine times, and we had many interesting discussions about our specific field of interest: Desert Botany, and a myriad of other subjects. In his old age he was a profound philosopher, who had read the works of Kant, Schopenhauer, Hume, and also of more modern thinkers, including Vaihinger, and the phenomenologists Husserl and Heidegger. Even on his last visit to Avdat at the age of 84, he often climbed the desert mountains like a chamois, often disappearing for hours on end. In the evenings, we sat around the kitchen table discussing epistemological problems, late into the night. I still recall the perplexed expression on our volunteers faces as they tried to understand what "those two old "exotics" were talking about!

As well as developing my skills as a physiologist and ecologist, I also had the opportunity to expand my horizons in other fields. An excursion into Bulgaria in the summer of 1932 brought me in contact with Balkan flora, which has many Oriental elements. This experience later helped me to understand the flora of the Middle East. The circumstances leading to this trip are an interesting story. Czar Boris III, the ruler of Bulgaria, was very interested in Botany and in a global list of botanists published at the time, his name appeared, his profession being listed as "King"! Boris Nikoloff, a student in Civil Engineering and Architecture at the Technical University, whose father was a colonel in the Bulgarian army and personal adjutant to the Czar, was a friend of Schrof, a former schoolmate of mine from Frankfurt. He arranged a meeting between me and Nikoloff, who on finding out my profession, suggested that we all visit Bulgaria. Shortly afterwards, we received a personal invitation from the Czar to visit his country and a permit to travel free-of-charge and first-class on the trains there. We gladly accepted this invitation, especially since Boris Nikoloff assured us that his father would make sure that we had everything we needed. Two architects from the Technical University, who were interested in studying the architecture of the old Bulgarian wooden houses, also decided to join us on the excursion.

On sending a copy of the King's letter to the head office of the Danube Steam-Ship Company, we received free tickets from Passau in

Germany to Lom in Bulgaria. We were not given cabins, but as it turned out, it was more pleasant to sleep on the open deck in the fresh night air rather than in the stuffy cabins.

In Passau, we bought enough food, tea and wine to last us for the whole trip, and packed it together with the various eating utensils in two large baskets. We got hot water for tea from the ship's kitchen. The trip took six days. On the way, we stopped at several ports: Linz, Vienna, Bratislava, Budapest, and Belgrade, which we managed to tour.

One of the highlights of the journey was travelling through the Iron Gates, where the Carpathian arc joins the Balkan mountains. Up river from the Iron Gates the Danube flows leisurely through the Hungarian plain, and broadens out into a "lake", while further downstream, it flows through a tremendously deep and narrow gorge, which is only about 300 feet wide at its narrowest point. There, on the southern side of the steep rock wall, I noticed holes hewn into the rock, remains of trails of the famous Via Trajana built by the Roman emperor Trajan (53-117 A.D.) This was part of a military road he built on the banks of the Danube in order to protect Moesia, one of the distant Roman provinces (today Rumania and Bulgaria) against constant attacks by the Barbarian tribes. The road had to go through the Iron Gates which were unnavigatable at the time due to the torrential rapids. Trajan overcame this obstacle by having large square holes made in the rock, into which thick beams, which were then covered by wooden planks, were driven. This road hanging in mid-air over the turbulent rapids of the Danube was a true masterpiece of Roman road-building. Only at the end of the last century the Iron Gates were made navigatable by the Hungarian government.

Years later, while driving along the banks of Lake Thun in Switzerland, I saw that a similar method had been used to build a road suspended above the lake. This serves as a good example of how a technique developed long ago, may be applied many centuries later, and adapted to suit construction methods.

I found the scenic beauty of the gently rounded hills crowned by castles and monasteries, wide seemingly endless plains, and the majestic Iron Gates breathtaking. But this was not the only attraction of the trip, it was also interesting to meet many different types of people. Our fellow passengers came from many countries: France, The United States, Bulgaria, Rumania, Yugoslavia, Albania, Macedonia, Hungaria and even one person from French Canada, whose old-fashioned French I could hardly understand. Most of them were students, and every day we invited someone else to join our small travelling party and to join us in meals, song, and general merrymaking. We also had a chance to discuss politics, and I learnt a great deal about the Balkan States. The "Macedonians", who lived in a region partly under Greek and partly under Yugoslav rule hated their respective "masters." Organized fighting, which today would be called terrorism, or freedom fighting, depending which side one supports, was widespread and condoned by the general public. One example of such hostility was the particularly volatile relationship between the Hungarians and the Serbs.

The root of all these conflicts could always be put down to a common denominator: a minority group, who was oppressed by the majority. In the Balkan states these were not allowed to teach in their native language in schools and in general, every effort was made to force them to assimilate into the dominant culture. The situation was further aggravated by the many different religious groups living in the area. Every country had Catholics, Greek Orthodox, Russian Orthodox and Muslim communities. I began to understand why the Balkan Wars of 1912, which I remembered from my childhood, were fought with such cruelty. Little had changed in the intervening years and the Balkans were still a keg of political dynamite. This enmity between the different national and religious groups frightened me, reminding me how far from mutual tolerance mankind was. I particularly felt the implications of this situation for the Jews, who were in the past always a minority.

The attitude of the many Bulgarians towards Hitler made me uneasy. On realizing that we were Germans, they always said: "Oh, you are Germans, you are our friends. How is Gitler (there is no letter "H" in Bulgarian), the saviour of Germany and Bulgaria?"

At Lom, the first stop in Bulgaria, we were received with open arms by Colonel Nikoloff, Boris' father, and we passed through the customs control without having to open a single suitcase. In contrast, all the others had to empty out their cases, and after inspection, each item was thrown onto one big pile, which every passenger then had to search through in order to claim his things. Everyone also went through a body search. This strict control at the port was the first indication that we were in a police state, and was reinforced many times when we passed through police or military roadblocks. As we discovered subsequently, these measures were designed to prevent Communist infiltration. In addition, all foreigners had to get special identity cards from the police, a procedure requiring about twenty different forms to be filled out and the submission of three photographs. Colonel Nikoloff took care of all this paper work, and we managed to get all the papers we needed in next to no time.

Shortly after our arrival, Colonel Nikoloff gave us a giant watermelon, the first I had ever eaten, and a basket of grapes. He then took us to a fine restaurant serving traditional Bulgarian cuisine, where we ate with voracious appetites since this was our first hot meal since we had left Passau! We then took a train to Sofia, where we had a special first-class luxury compartment, the first, and also last time, that I ever travelled in such luxury and comfort!

Upon arrival in Sofia, we visited the King, who granted us a short audience in his palace. He told his First Secretary to help us to finalize the plans for our trip, giving us written recommendations to officials in the various places on our itinerary. The King also invited Nikolai Stojanoff, Professor of Botany at the University of Sofia, to that meeting and instructed him to help me with my botanical work. Stajanoff, a Taxonomist and Plant Geographer, was considered to be one of the leading experts on the local flora.

Professor Stojanoff suggested that we begin with a two-day expedition of the Vitosha, a mountain massive a few kilometers south of Sofia. During the excursion, he introduced me to the flora of the country. It is to his credit, that I was able to understand the complicated Plant Geography of Bulgaria. Later he also helped me to identify the plants I had collected.

After our returning from Vitosha, we took a train from Sofia to Vrana, where there was a 900 hectare Botanical Garden established by Czar Ferdinand, Czar Boris' father. This park was one of the most interesting and impressive of its kind that I have ever seen. Czar Ferdinand's ambition was to collect all Bulgaria's flora there including the many endemic species, and he had succeeded remarkably with his task.

The Alpine Flower collections each representing the plants found in the Bulgarian mountain massive were the most interesting part of the Botanical Gardens. Special emphasis was placed on endemic and rare species. The Park also had excellent specimens of the indigenous trees in Bulgaria.

The Director of the Garden, a German, was an excellent professional gardener, as well as being a good botanist. I spent several hours with him and he gave me a guided botanical tour, that proved to be invaluable to me when I later ventured out alone.

The tour of the Botanical Gardens lasted so long that special transport had to be arranged to take me to the station. One of the

King's chauffeurs came to pick me up in an official car. When we arrived at the station the train was already pulling away, but when the engineer spotted the King's coat of arms on the car, he stopped the train immediately, and got out of his cabin to escort me to the first-class luxury compartment, usually reserved for the King. All the passengers looked out of their windows to see who this "dignitary" was. I noticed that some of their looks were unfriendly and hostile, which I later understood could be put down to the deep political tensions dividing the country.

After this general introduction to Bulgarian flora, we went all over the country, from South to North, East to West, and finally back to Sofia. On our trip we went to the Rila and Pirin mountains, the Rhodope range, Plovdiv, Tatar Pozardjik, Drenovo, Tirnovo, Kazanlik, Shoumen, Burgas, Varna, Pleven and then returned to Sofia. Climbing the high mountains was the most exciting part of the trip. We went up to the Tcherni Vrach (2286 m), in the Pirins, to the Vikhren (2915m), and in the Rhodopes to the Mussala (2925m) the highest point in the Balkans. The wild and the yet untouched virginal splendor of these mountains impressed me as much as the beauty of the flora.

Bulgarian mountain flora is unique in that it contains a large number of endemic species. Stojanoff had told me that about 27% of all the higher plants in Bulgaria are endemic, and that most of these species occur in the high mountains. The "resurrection" plant *Haberlea rhodopensis* I saw in the Rhodope massive is a good example of an endemic species. The only other related species known is found in China. This is an extreme case of related species that inhabit non-continuous (or disjunct) areas. As far as I know, there is only one possible explanation for this phenomenon. In the past, under different prevailing conditions from those of today, these two species probably had a common ancestor which grew in a extensive contiguous area. As a result of climatic or topographical changes, these species disappeared except in a few limited regions where they remain as "relics."

The *Haberlea* is not only interesting as an example of a plant inhabiting a disjunct area, but it also exhibits strange physiological behavior. It can be desiccated and kept in this state for a long period. When rehydrated, it continues growing as if nothing has happened, hence, the name, resurrection plant. Most higher plants die after such treatment. The few known species of resurrection plants display behavior typical of lichens, mosses and some ferns. Is their ability to "resurrect" a genetic trait that has been preserved during the course of evolution, or is it a relatively recent evolutionary "re-discovery"?

Finding the *Haberlea* and other endemic species was not less exciting than the fact that many places I "botanized" were botanical no-man's land. Stojanoff later pronounced that several plants I had collected at the highest points in the Rhodopes were new species for Bulgaria, and others to the world, which made me very proud of myself.

Exploring for interesting plants was not the only thrilling experience. The flora in the mountains of Bulgaria represents a meeting point of plant elements, whose center of distribution extends far beyond the country's borders. This experience was like a "living textbook" in Plant Geography. I came across the common beech and the ash, both Middle European elements that I knew well from Germany, as well as the pine tree, a boreal element, which covers Northern Europe to Siberia, and whose southernmost point of distribution is in Bulgaria. I also saw the Judas Tree and the Kermess Oak, both Mediterranean elements, which are common in Palestine, Irano-Touranian plants whose center of distribution is in the Central-Asian highlands, and Pontic plants, typical of the Black Sea coast. There were also various endemic Bulgarian elements, such as the Moesian beech and several species of oak.

Although these forests fascinated me on the scientific level, their emotional impact went even deeper. The forests of the high Rhodopes and the Pirin were dense, primordial creations of nature undisturbed by man, in which individual trees were majestic giants. I had never before seen a primeval pristine forest, which had once completely covered Europe. Owing to man's "progress", only a few isolated areas like this remained at the time, and I doubt whether the Bulgarian forests we traversed in 1932 are still as pristine now as they were at the time. We saw already too many people cutting down the largest trees. An entry in my travel journal, written one night in a "Caruza" (a small horse-drawn vehicle) en route from the Rhodopes to Plovdiv, summarizes these impressions: "Rising full moon, clear star-spangled sky, driving at a terrifying pace over rocks and boulders, deep gorges, torrential brooks, through endless forests, giant trees, archetypical trees, silhouetted against the starlit sky, I am enraptured by the primeval beauty of these forest and trees that go back to the dawn of creation."

It was not always easy to go into the mountain ranges, which were far off the beaten track and with no proper roads. For example, for the three-day journey to the Mussala, we had to hire a mule driver, who showed us the road and carried our provisions to a mountain hut at the foot of the summit, which we climbed later. That night we slept in the hut, and I had a very bad attack of dysentery, forcing me to go out of the hut although it was bitterly cold. By the morning, I was too weak to stand on my feet and could not walk. In order to climb back down the mountain, I attached myself to the mule's tail, tottered after him down the steep mountain path. Shortly afterwards, I could not go on any further, I felt as if I was dying, and told my colleagues to abandon me there. The mule driver responded by taking a huge onion out of his pack, and told me to eat it all: "This old Bulgarian folk medicine against dysentery always helps." With great revulsion, I forced the onion down my throat, fighting the urge to vomit. After resting for an hour, the dysentery was gone, and I could walk again.

Although the mountains and their flora were the most significant part of the trip for me, the old Bulgarian wooden houses fascinated my colleagues. No one had ever studied these before, and there were enough interesting houses in the villages of Trevna (Tryavna), Drnovo (Dragonovo) and Tirnovo (Turnovo) to conduct a serious study. We walked through the narrow streets of rugged cobblestones, looking for the most impressive wooden houses. When we found a house to study, it was photographed from all angles, and then surveyed until all the cross sectional and floor plan features were clear. I participated in and enjoyed this work. As I wrote in my journal: "The most remarkable aspect of these houses are two or three floors on heavy logs that jut out of the building. The facades are divided into a number of alcoves, all natural wood, which over the years has become a deep brown color. Many houses have porches enclosed by solid wooden walls, which are like rooms, but open onto the street, their low ceilings giving a feeling of coziness. The roofs with double cornices on large protruding logs covered by wooden shingles held by large irregular unhewn sandstone blocks to provide weather protection, make the building harmonize with the surrounding. The richer homes have wooden hand carved ceilings in geometrical arabesque patterns."

While the flora and local architecture of Bulgaria were the central points of interest, we also saw its other fascinating sights. We visited the Rila and Bathchcovo monasteries with their beautiful old frescos and iconostases. We also saw many other churches and monasteries built in the Greek Orthodox style, monuments to the past splendor of the Byzantine empire. I also came across a mosque and heard a muezzin's call for the first time. I became interested in the history, language, politics and people of Bulgaria. In my ignorance, the only thing I really knew was that the country had been under Roman, and then Byzantine rule. There seemed to be a great deal of political tension among the different ethnic and national groups in that area, including the Turks, Pamacs (Bulgarian Muslims living mainly in the Rhodopes), Macedonians, and Karakatchans (gypsies who spoke a Roman language).

Since the Bulagarians speak a Slavic language similar to Russian I naively assumed that they were Slavs. This is certainly not true. At the time of the occupation by the Romans, Moesia, which now is called Bulgaria, was inhabited by Thraco-Illyrians. It remained under Roman rule for 500 years and their influence is still felt today. For example, Plovdiv which was once called Philipopolis, was founded by the emperor Diocletian (50-96 A.D.) From the 4th century onward, Slavic tribes from the East began to infiltrate Moesia and subsequently occupied the area. But the real Bulgarians who were Turanians from the southern region of the Volga closely related to the Huns only came to Bulgaria at the end of the 7th century. They changed the name Moesia to Bulgaria, and adopted the language of the Slavic population living there. In time, the Turanians, Slavs, and the Thraco-Illyrians became integrated to form the Bulgarian people of today.

Bulgaria became a Christian State under Byzantine rule in 870 A.D., the Emperor Czar Boris I accepting Greek Orthodox Christianity. For this reason, there are many Greek Orthodox churches and monasteries in the country.

In 1396, the Turks conquered Bulgaria, and for 500 years it remained a Turkish province. The Bulgarian population then revolted against these cruel rulers, and the Russians intervened to liberate their "Slavic" brothers. Without their help, the bloody revolt with its decisive battle of the Shipka pass in 1877, would have certainly failed.

Most Bulgarians are strongly pro-Russian. Only among the elite, ruling class, whom we got to know through Colonel Nikoloff, was love of Russia mixed with a fear of Bolshevic communism. This feeling explained the ubiquitous military and police presence.

One evening, while walking alone along the streets of Tirnovo, I came across a group of people standing around two Austrians playing guitar and a very attractive German girl with two long blond plaits. The group sang German songs, and after each song the girl collected money from the audience. I talked to them and followed them from one coffee house to another as they performed. During the course of conversation, they professed to being communists. Gertrude von M. from Hamburg told me things about Bulgaria that I not heard about before. She mentioned the "Green Communist International," an agrarian communist movement, which had many partisans and sympathizers among the Bulgarian peasantry. In 1923, this organization had tried to overthrow the government, and the revolt was suppressed with much bloodshed. On my return to Darmstadt, I verified her story.

They were very interested in this movement, since, in contrast to Russia, peasants, and not the factory workers, formed its backbone. Later, in very cautious conversations with several Bulgarians, I found out that even in 1932, this form of agrarian communism was very much alive. The government did everything in its power to fight it, including the introduction of compulsory labor service (Trudovacs), an idea later adopted by Hitler in his "Arbeitsdienst" (Labor Service Corps).

In Bulgaria I also came into contact with Sephardic Jews (of Oriental origin) for the first time. They were descended from Spanish Jews, who refused to convert to Christianity in 1492, and were consequently expelled from Spain. They fled to the countries of the Orient and the Middle East, including Bulgaria. Five hundred years after their expulsion, they still spoke the 15th-century Spanish-Hebrew dialect, Ladino. In Sofia, I had the honor of meeting the *Sephardic* Chief Rabbi, who sang some long forgotten old Spanish ballads to me one Saturday afternoon. Many years later, as Vice President of the Hebrew University, I entertained a group of Spanish Jesuits, who had come to Israel in order to collect these ballads; their anthology was published in a journal, "Spharad" (Hebrew for Spain).

In autumn 1932, we returned to Darmstadt. Open anti-semitism in the streets had increased considerably. Large groups of SA men paraded through the city singing lustily: "Wenns Judenblut vom Messer spritzt, dann geht's nochmal so gut" (When Jewish blood spurts from the knife, things go twice as well), to wild applause from many of the passers-by. My wife Alice once came to see me at the university. In tears, she told me that while taking our dog for a walk in a small park, a woman came up to her and shouted: "You Jewish sow, you should be forbidden to keep dogs. They are too good for you dirty Jews!"

My belief that the Germans were already lost to a blind and irrational anti-semitism was also reinforced by another incident. On a train ride to Darmstadt from Frankfurt, I and several others shared a compartment with a high-ranking SA officer. For reasons unknown to me, he suddenly began the most vile anti-semitic tirade. The other passengers pretended not to hear him. After a while, I could not stand it any longer, and said: "How can you vilify the Jews, when your own Savior, Jesus Christ was a pure-blooded Jew?" The officer nearly jumped out of his seat, screaming: "That is a damned Jewish lie!" Jesus was a blond blue eyed Aryan!" I replied: "Do you really believe all the rubbish your Führer tells you?" Although many of those in the compartment smiled to themselves, none had the courage to say a word. Fortunately for me, we arrived at the Darmstadt station a few minutes later, where the officer was greeted by a horde of SA men.

On April 1, 1933, a day that will never be forgotten by those Jews that lived through it, the well-known boycott was carried out against all Jewish owned stores. That morning I went into town and saw several SA and SS men standing in front of every Jewish shop, telling all prospective customers not to go into them, since they belonged to "Drecksjuden" (dirty Jews). They said that true Germans would not buy anything from these "Untermenschen" (subhumans). People who ignored these warnings, were brutally beaten. For good measure, the SA and SS men also beat up Jewish looking passers-by. The Nazis pasted up large signs reading: "I am a dirty Jew, Aryans will not buy from me" on the shop windows. Uniformed Nazis on the street corners rattled charity boxes shouting: "Give money for one way tickets to Jerusalem for the Jews." I could not stop myself from walking up to one of them and saying: "Thank you for collecting money for me. I will make good use of it soon, your offer of a ticket to Jerusalem is much appreciated." Before he realized what I said, I disappeared into the crowd. When I got back to the university and told Huber what was happening in the streets, as a devout Catholic, he could not believe it. He said that the "good" Germans in the crowd would never tolerate such excesses. This gave me a revealing insight into how politically naive people like him were, unknowingly, they contributed to Hitler's rise to power.

Just as I finished speaking to Huber, the telephone rang. The rector of the university summoned me immediately to his office. With difficulty, he said, "Herr Doktor, you have been denounced as a confirmed Jew (bewusster Jude) and I am supposed to dismiss you immediately. However, I am willing to give you four weeks to leave the university." I had expected something like this and answered rather brashly, "Your Magnificence (the official title of a rector), in time to come, you may remember this day as the beginning of the downfall of Germany. As for the four week's grace, you can keep them, I am leaving this afternoon." With these words I got up, and walked out banging the door. I immediately sent a cable to the Aaronsohn family telling them to expect me in Palestine in April instead of October. Walking out of the rector's office, I met a Jewish colleague and assistant from one of the technical departments in the university. I told him what had just happened to me. He replied: "You see, that is what comes of always telling everybody that you are a Jew and a Zionist. It could not happen to me." How wrong he was!

In the Department, everyone already knew what had happened. I immediately started packing everything I wanted to take with me to Palestine, including a torsion balance for assessing transpiration and Huber's apparatus for measuring photosynthesis in the field. In 1932, when I first revealed my plans to go to Palestine to Huber, he applied to the "Notgemeinschaft der Deutschen Wissenschaft" (an organization that gave grants to scientists) on my behalf. Although promised funds, the grant was immediately revoked upon my dismissal from the university. My father stepped in and paid for the instruments, and I took them with me to Palestine.

While helping me to pack, my students told me they were sorry to see me go, but pleased that I had been able to leave for Palestine in time. I was very surprised that they were not wearing the Nazi party emblem on their lapels and when I questioned them about this, they answered me in their heavy Darmstadt slang: "Herr Doktor, today we do not want to be identified with the rabble wearing our emblem", hinting at the fact that on that particular day, April 1, 1933, many people had suddenly discovered that it was adventageous to be a Nazi. With the exception of a few books, we left everything in the apartment intact, including the furniture, and did not lock the door. Presumably, the belongings we left behind were taken by who ever got there first.

One of my students called his father, a professional mover, and he came with his van and transported us together with our luggage to my parents' flat in Frankfurt. He refused my offer of payment. Three weeks later, we arrived in Haifa.

I cannot close this chapter describing my personal experiences of Nazism without emphasizing how systematic repetition turned a single word into a deadly weapon. From the very start of his political activity, Hitler and later, his henchmen including Goebbels, Streicher, and Sauckel characterized Jews and other "inferior" races as "subhuman". Before his rise to power, the constant use of this term was not recognized as a psychologically manipulative measure which would ultimately allow him to carry out unthinkable crimes against the Jews by pointing out that they were "subhuman". As such Jews were excluded from public office (I may have been one of the first to be dismissed), civil service, and the practice of medicine and law. This measure was followed by the so-called "Nuremberg Laws", compulsory confinement in ghettoes, and finally, the Holocaust.

Since Jews were "subhuman", the majority of the German people agreed with Hitler that Jews had to be treated like dirty scum by definition, they were not human. Regrettably, Hitler is not the only example of a man who used language as an instrument of murder. As Matthews (15:11) succinctly points out: "Not that which goes into the mouth defiles a man, but that which cometh out of the mouth this defiles a man."

Chapter 3
Palestine

The Aaronsohns and Oppenheimer received us with open arms, and had already rented a flat in Jerusalem for us. Over the next few years, I spent several days a month in Zikhron, where the Aaronsohns lived, travelling there by train. I worked in the herbarium there, taking the plants I could not identify back to Jerusalem, where I took advantage of the facilities at the Hebrew University's Department of Botany. However, working with dried plants did not satisfy me and I wanted to get to know the live flora of the country. At every opportunity, I searched for plants I had observed in the herbarium in their natural environments. Oppenheimer and I published the results of our studies in "Florula Cisjordanica" (15). Much later, when I started my work in the Negev, the knowledge I had gained on the flora of Palestine came in very useful.

A few days after arriving in Jerusalem, I was introduced to Alexander Eig, Head of the Hebrew Univerisity's Department of Botany by Oppenheimer, who directed the Section of Plant Physiology and Anatomy there. Both the Department and the University as a whole were in their infancy at that time.

Although the foundation stone was laid in 1918, shortly after the British occupation of Jerusalem, the University only opened in 1925. Initially, it was a research institute, and opened its door to students two years later. Mount Scopus, the site of the University, has an incredible view. To the east lies the yellow-brown Judean desert and the Dead Sea with the Moab mountains shrouded in mist in the distance, and to the west is the old city of Jerusalem, the Temple Mount, Omar and El Aksa mosques, and the ancient city walls almost seem to be within reach. At sundown, the dome of the mosque of Omar glows like fine gold in the sun's dying rays.

Otto Warburg, scion of a renowned Jewish family, who was a botanist of international standing, founded the University's Department of Botany. His three-volume work, "die Pflanzenwelt" (The World of Plants), became a standard textbook; and from 1911 to 1919 he served as president of the Zionist organization, showing a particular interest in agriculture and botany. At the turn of the century, he got to know Aaronsohn and helped him in many ways.

In 1921, Warburg became Director of the Zionist organization's Agricultural Experimental station, which was first established in Tel-Aviv and later transferred its activities to Rehovoth. With the establishment of the University in 1925, he was appointed as the first professor in the Department of Botany, which was initially a joint enterprise of the Hebrew University and the Agricultural Research Station at Rehovoth. The Department as an independent unit moved to Jerusalem in 1931.

Since 1925 Alexander Eig had served as Warburg's Assistant. A Russian Jew who had come to Palestine in 1909, he had collected a large herbarium of local flora. Shortly afterwards, Michael Zohary and Naomi Feinbrunn joined the Institute. This trio, Eig, Zohary and Feinbrunn, collected plants in Turkey, Lebanon, Syria and Iraq, and turned the University's herbarium into a central collection covering the entire Middle East. Although they always maintained their interest in Taxonomy, they became increasingly attracted to the Geography and Sociology of the Plant World as a result of the many field trips they made.

In 1933 Warburg died, and Eig was appointed as head of the Department. Oppenheimer left the Department at that time to take up the post as Head of the Department for Citrus Research in the Agricultural Research Station in Rehovoth. I replaced him at the Department of Botany in 1934.

A great deal had happened in the meantime. Shortly after our arrival, Oppenheimer introduced me to Richard Richter, who was a very versatile and skillful man. During the First World War, he had been a fighter pilot in the German Richthofen squadron. After the war he emigrated to Palestine with a group of "Chalutzim" (Jewish pioneers), went to live on a kibbutz for a short time, took part in various archaeological excavations, and then became a tourist guide. We immediately became friends. I asked him to help me to start my eco-physiological studies of the Judean desert, because of his knowledge of the country. I had brought the equipment that Huber had developed for such investigations with me from Germany. Richter and I reconnoitered the Judean desert to find a suitable location for our work. Since we could not effort to hire a car, we chose a site close to the main road. We found a good place, which we called "Wadi Rotem" because there was a large shrub of the desert broom (Rotem) growing there. We began our study during a *Chamssin*, which is characterized by dry desert winds and extremely high temperatures. By noon it was 42°C, and being unaccustomed to such heat, it felt like being in a furnace. In the evening, we had to carry our heavy equipment back to the nearest stop of the Arab bus line on the Jericho-Jerusalem route. On the way Richter collapsed under the heavy burden, but fortunately recovered after a while. We used up all our strength to get to the bus stop, and arrived in Jerusalem half-dead.

We subsequently spent many days and nights in *Wadi Rotem.* But it soon became clear that the equipment I had would not be sufficient. Since we could not afford to buy expensive new instruments, we had to improvise using the primitive means available. In this respect, Richter's skills came in very handy, for example, he made a drying oven needed for determining the dry weight of plants and soil, which was maintained at a constant temperature with a "thermostatic" electric bulb. In 1937 after three years work in the desert, we published the results of our research (24).

Stocker's classical book (23) on the ecophysiology of Egyptian desert plants had whetted my appetite, and through my work in the Judean Wilderness, the desert environment enchanted me more and more when I visited many other deserts in Iraq, Jordan, Egypt, North America, Mexico and Australia. It is difficult to explain my emotional reaction to the desert.I was bewitched by the shiny regs (plains covered by black pebbles), stretching as far as the eye could see, the steep multicolored mountains, the rocks eroded into fantastic shapes, and the deep gorges with red, yellow and white walls of Nubian sandstone. The nights were crystal clear, the glittering stars in the sky seeming within close reach. I was spellbound by the almost audible silence, in which the prophet Elijah heard God's voice according to the Bible (I Kings 19: 11-13): "....and a great and strong wind rent the mountains......but the Lord was not in the wind, and after the wind an earthquake; but the Lord was not in the earthquake; and after the earthquake a fire; but the Lord was not in the fire; and after the fire a small, still voice. And it was so when Elijah heard it, that he wrapped his face in his mantle." I can still hear this still voice and can understand why God spoke to Moses, Jesus and Mohammed in the desert. I understood why the idea of the one and only God was born in the desert. For in this harsh monolithic environment, where soft and tender thoughts cannot divert the mind, one feels small, insignificant, and lonely within the infinite Universe; only the presence of HIM that is felt with all the senses can give man confidence and peace of mind and spirit.

The desert is also unique in scientific terms, since in contrast to the diverse external conditions affecting life in other environments, one can think of jungles, forests, or even lawns, in the desert lack of water, is the single dominant factor to which life had to adapt.

The year 1933 was critical for my future in many ways - I left Germany, came to Palestine, began research on desert environments and in August of that year Eig asked me to join an expedition to the forests of Kurdistan. Our Department had been invited by the Ministry of Agriculture of Iraq, which had only become independent in 1928, to assess the size, make a list of the kinds of trees and plants, and to investigate the ecology of the area. Since these were virtually the only forests in Iraq, they were an important national asset. We were also asked to suggest measures to protect the forests from ruthless over-exploitation by the local population.

Eig delegated the responsibility of investigating the eco-physiology of the trees and testing the soil types to me. We packed the equipment for a mobile laboratory in crates. At the end of our expedition we prepared a comprehensive report for the Iraqi government emphasizing the importance of forest protection laws. We even suggested that they adopt the "New Year For Trees" - schoolchildren go out to plant trees on this day - that the Zionist organization has successfully introduced in Palestine. We also made a complete list of the plants we had found in the forests which was subsequently published by the Iraqi government. Unfortunately the political changes that have taken place since 1933 and the war between Iraq and the Kurds have made it impossible to envisage similar cooperative projects at least in the foreseeable future!

August 31, 1933, the team, including Eig, Zohary, Feinbrunn, Duvdevani, the gardener Amdursky, the geologist Pickard and I left Palestine in two overloaded desert cars. We first traveled to Amman in Transjordan, which was ruled by Emir Abdallah at that time. I was particularly impressed by the magnificent Roman, Byzantine, and Nabataean ruins, without suspecting that in years to come, the farming techniques used in these cultures would provide an important source of inspiration for my scientific work.

We then drove through the Jordanian-Syrian-Iraqian desert, which is part of the Great Arabian desert. There, I had my first experience of a mirage. At each stop we collected plants for our herbarium, often working into the small hours of the morning. In Kasr-el-Azraq, on the border between Jordan and Iraq, I wandered off alone and came across a large Jordanian fortress built along the lines of the French installations in the Sahara desert, which was guarded by a Bedouin corps. I used my newly-acquired Arabic to speak to the young

My mother and father in front of our house in Jerusalem

Ben Gurion (*right*) in his "cage" during the opening of the Givath Ram campus in April, 1958 (Photo Hebrew University)

Shmuel Duvdevani in our "lab" in Kurdestan (Photo M. Evenari)

The author and his wife Liselotte in the ruins of Shivta (Photo N. Tadmor)

legionnares who made an impressive sight in their white Abbayas, red-white Kefiyas, and their leather belts bearing shabrias (daggers) in beautiful silver sheaths. They became very friendly, taking me by the hand and leading me into the fortress, where we drank Hel (cardamon)-flavored coffee in finjans (small cups). I began to feel increasingly uneasy when they came nearer and nearer and started touching me all over. Fortunately for me my colleagues rescued me in time from being raped.

By the time we arrived in Baghdad, we were exhausted. I found the town disappointing since I had a picture of the fabulous town of Harun al Rashid in my mind. We then made our way by train to Kirkuk. There we saw a sea of derricks pumping oil out of the ground. The pipeline, which later carried crude oil to the refinery in Haifa, was in the initial stages of construction. Seeing this enormous new industry I wrote in my diary: "I do not believe that these new riches can be a blessing either for the young state of Iraq or for the world as a whole. They will undoubtedly become a focal point of worldwide political intrigue." A similar opinion was voiced by a high official in Baghdad, who considered these large oilfields as a gift from the devil. In the distance, we could see the faint outlines of the mountains of Kurdistan. We then went by car to Suleimaniyeh at the foot of the Kurdish mountains, from where we had to proceed on horseback. Our police escort let us choose horses for the ride. Whereas my colleagues chose the tamest ones, I, driven by ambition, selected a fiery stallion. Never before had I mounted a horse but wanted to show our police escort that a Jew could ride as well as they. I soon learnt that riding is an art that has to be mastered.

The next day, at 5 o'clock in the morning, the local people came to see off our strange caravan, consisting of seven expedition members, an Iraqi Ministry of Agriculture official, 19 local policemen, and 27 pack-horses and mules with their Kurdish drivers. While on flat ground, my horse trotted slowly, but when one of the mounted policemen overtook me, my horse broke into a fast gallop. The policeman must have thought that I wanted to have a race with him. Both horses ran faster and faster, and I had no idea how to hold mine back although I was terrified of losing control and falling off. Covered in cold sweat, I bent forward and grasped the horse's neck with both hands, unknowingly adopting the posture of a jockey during the final stages of a race. The policeman followed suit. Miraculously, my horse won the race, firmly establishing my reputation as an excellent horseman.

Fortunately for me, further opportunities for races did not arise, since we had to go very slowly in the forest. For weeks on end, we rode through the picturesque forests and mountains of Kurdistan along the Iraqi-Irani-Turkish border - from Suleimaniyeh to Amadiyeh, Zakho, Rawanduz, Dihok, and Mosul. Already on the first day, we passed through many of the lower mountain ridges, pressing on at nightfall. The light of the full moon gave the silent landscape a fairy-tale quality. As we went through sleeping villages, the Kurds rode at the head of the column, singing their strange songs. At two in the morning, we reached the village of Jaafaran, where for several hours, we dozed rather than slept on the village green. In the morning, on the advice of the sheikh and some of the villagers, regarding the sites of the largest forests, we went on to our first stopping place, Kamitachat, where we spent several days. This proved to be an ideal place for our work as it was deep inside a large forest, and had a source of clear potable water. We were surrounded by trees, oaks dominating with small stands of wild pear trees, maples, ashes and pistachios.

We set up four tents in Kamitachat. Mine was in the shade of a giant willow tree. Next to our tents the Kurds put up a mobile kitchen and built a "laboratory" out of branches. Our working day was well organized. Every day at dawn, Eig, Zohary and Feinbrunn went off to map the forest. One of us always remained behind in the camp. Two Kurds and two policemen accompanied me to carry out eco-physiological measurements of the trees water balance. Duvdevani worked in the makeshift laboratory testing soil samples. At nightfall, we would all get together again for the evening meal, consisting of chicken prepared by our cook Abba. Afterwards, we put the plants that we had collected during the day into the herbarium, and held leisurely discussions of the day's events. It was a paradisic life.

During this period, I got to know the Kurds as they worked alongside us. I tried to learn their language so as to be able to communicate with them. They showed a remarkable knowledge of the plants growing in their country. Once, they brought me two plants that looked very similar, which they referred to by two different names. On further investigation, Eig found but that these really were two different species. The Kurds soon struck up friendships with me since I served as the expedition's physician and pharmacist. Everyday about 20 Kurds from the surrounding areas would come and see the *"Hakim"* (physician). One day a policeman brought an old Kurd to see me, explaining in his broken English: "This man ten day no go." I gave him a powerful laxative, which produced immediate results, and greatly boosted my prestige. Later in Rajat, a village which is situated at a height of about 4000 m, I massaged the body of an old sheikh who could not walk properly. When this treatment proved successful, I was "promoted" from "Hakim" to "Hakim Basha" (chief physician) by his tribe.

From observing the Kurds at close hand, it was obvious that they were not of Semite origin. They are very different from the Arabs, whom they hate and fight to this day. Their clothing is distinctive, and they have blue eyes and brown or blond hair. I could tell that their language belongs to the Indo-Germanic group from the numbers, which sounded familiar to me. Each village spoke its own dialect so that, for example, our cook Abba, hardly understood the Kurds of Amadiyah.

We have very little knowledge of the historical background of the Kurds, but it is well-known that they are a belligerent nation. Under Turkish rule, they rose against the government many times. Later, led by Sheikh Mahmud, ruler of Suleimaniyeh, they fought both the British and the Arabs. The Arab official in our team was frightened of them and so he slept with a loaded pistol under his pillow. One of the Kurds, on being asked how we would have fared without our escort, gave us to understand by his hand movements that our throats would have been slit! They often mentioned their firm desire to become independent - very little has changed in this respect since 1933.

The villages of the Kurds blend in so well with the landscape that from a distance they seem to vanish into the surroundings. Their houses, made out of loam, have square patios, where the wealthy have constructed arcades supported by wooden columns. According to one view, the Kurdish house is the prototype of the ancient Greek ones.

The Kurdish villagers living in the fertile areas are farmers; while those inhabiting the mountain regions are semi-nomads. The cattle are kept in the villages during the winter and are taken into the mountains in summer. The forest is one of their major resources since they export valuable oak timber. In addition in the fall, using long sticks, they gather galls from the oak trees, which contain tannic acid that is much in demand in European tanneries, and accounted for an even greater share of the exports than timber at that time.

The oaks also provided the Kurds with manna. At the end of the growing season, oak leaves are covered by manna, a layer of a sugary material, excreted by aphids when they feed on the leaf sap. The peasants collect these leaves and soak them in water, which is then boiled in large cauldrons until it turns into syrup. They either spread this syrup on bread, or use it to make "Turkish delight," which I remember buying as a child at fairgrounds.

We stayed in Kamitachat for 10 days, and then went on to Svaratuka near Suleimaniyeh, where we worked in the dense oak forests. Then we visited Amadiyeh, a picturesque town on a high plateau surrounded by mountains. The plain below, where the residents had their "summer huts", was a huge garden laden with vegetables and fruit trees.

Mosul, another town we visited, had a mixed population of Kurds, Assyrians, Chaldaeans and Jews. The Kurds were reputed to be rather aggressive, as we soon found out for ourselves when we passed several villages that had been left to waste. From our guide, we learnt that the Assyrians there had been massacred by the Kurds. This was subsequently confirmed by a survivor with whom we communicated quite well since the Assyrians, who are Christians, speak an Aramaic dialect resembling the language of the Talmud. An example: one of their prayers is "Abuna deschamaya, lachman sukanan hav lan....." (Our father in heaven, give us our daily bread), which we could understand word by word. He told us of the long history of tension. However, the Kurds had been kept at bay while Iraq was under British mandate, but in 1932, when Iraq became an independent state, they mercilessly attacked the Assyrian villages, plundering the homes, raping the women, and murdering most of the inhabitants. They occupied or razed the villages to the ground. The few survivors fled in fear of their lives.

The Assyrian patriarch appealed for help to the League of Nations, but only received words of sympathy. We were confronted with a classical example of a defenseless minority that did not get any outside help. Why did the Christian world ignore the plight of their brethren?

As Jews in Palestine, this lesson seemed particularly pertinent - we would only survive if we could defend ourselves. Indeed, later, during the Holocaust nobody came to the defense of the Jews, and only when it was too late, the world claimed to be "sorry" about the tragedy that had taken place.

The Jews of Kurdistan, with their ancient roots and unusual living conditions merit a special chapter. Later they settled en masse in Israel. I came across Jews form the mountain region for the first time in Amadiyeh. There, in one of the shops in the Bazaar, I bought a ring. I noticed that, unlike the Kurds, the shop owner had a beard and looked Jewish. I started speaking to him in Hebrew, which he understood. He told me that he was the *Chacham* (President) of the local Jewish community and invited us to his house. In contrast to the Kurds, the Jewish men carried no weapons. "This is a sign of submission to the Kurds. If we were to carry weapons, as non-believers in Islam, they would probably put us to death", the Chacham explained. The women were shy, rarely went out in public, would not talk to strangers, and usually refused to allow their photographs to be taken.

According to tradition, Jews had lived in Kurdistan since the first exile from the Land of Israel when the Jews of Samaria were carried off in captivity to Assyria in 720 B.C. Apparently Jews remained there even though the Persian King Cyrus (587-529 B.C.) made a decree allowing Jews to return to the Holy Land.

The Jews living in Suleimaniyeh, Amadiyeh, Zakho and Mossul were artisans, merchants and peddlers; and all the silversmiths and weavers in these towns were Jews. Although they prayed in Hebrew, they spoke Aramaic at home. They engaged in study of the Talmud and kept the religious practices. They were afraid of the Kurds and felt particularly insecure after they saw what happened to the Assyrians. They asked many questions about the condition of the Jews in Palestine as they wanted to go and live there. Interestingly in Motza, near Jerusalem, where we now live, the old-timers in the settlement of Jews of Kurdish origin, vividly remind me of that time.

We went to pray in the synagogues in Suleimaniyeh, Amadiyeh and Zakho. These were build in the same style as their homes - a wooden door led on to a square patio where the men prayed sitting on cushions, under the open sky in the summer. The *Torah* scrolls were placed on the central wooden lectern. The large room on one side accommodated the Holy Ark, where the *Torah* scrolls were stored, and was used for prayers during the winter. The larger synagogues in Suleimaniyeh and Zakho also had covered arcades supported by wooden columns around the patios. The Hebrew inscriptions carved on the capitals were so old that many of the letters had worn away.

On Yom Kippur, the Day of Atonement we prayed in the synagogue at Zakho. As we came in all the people rose to their feet and shouted "Shalom." We were given the honor of reading from the Torah. As European Jews, it felt strange to sit cross legged among these "Kurdish-looking" men. The women, seated on the roof above, wearing wide Kurdish breeches and colorful overgarments, with their faces and hands painted blue, rings through their noses, bracelets round their feet, and wearing giant turbans crowned by multi-colored feathers, were a feast for the eyes!

On *Rosh Hashana* (the Jewish New Year), although we arrived at the synagogue in Mossul toward the end of the service, we were greeted like "Messiahs". They blew the *shofar* (rams horn) again specially for us.

The profoundest impression of our whole expedition was to encounter Jews living in small villages. The *chacham* of Amadiyeh told us about Sandor, a small village nearby where all the inhabitants were Jews. We immediately visited the village. The loam houses there were surrounded by poplars, vineyards, orchards and fields. The first people we came across were so frightened of us that they did not recognize us as Jews. Our encounter with a weaver, who was spinning yarn on his balcony, was more fruitful. He led us - together with our Kurdish escort - into a guestroom of the chacham. He then brought over the *shamash* (beadle) of the synagogue who cautiously started to talk to us in Hebrew.

However, we noticed that he only gave brief answers to our questions, and would not let other Jews talk to us. Only after our Kurdish escort left did the chacham open up, explaining that since we were Europeans and accompanied by Kurds, he had initially suspected us of being spies. Now that the Kurds had gone and he was certain that we were *bonafide* Jews from Jerusalem and had found out what we were doing in Kurdistan, he could speak freely. He told us that after the Assyrian massacre the lives of the Jews in Sandor had been in danger, and many people had fled to Mossul. They were overjoyed to see us and asked about relatives who were already living in Palestine. They told us that they wanted to come to the Holy Land and asked whether we could help them to realize this dream. After our return to Palestine, we reported our experiences in Sandor to Ben-Zvi, who was President of the Jewish Agency at that time and later became the second President of the State of Israel. He had not previously known of the existence of this Jewish village, but promised to do all that was within his power to help. Now, all the Jews of Sandor live in Israel.

Sandor was an eye-opener for us - a village inhabited only by Jews, who for as far back as they could remember, had been peasants who made a living by their agriculture selling produce and hand-woven carpets that were colored with natural dyes derived from plants. We visited their well kept vineyards and fields. These Jews had been peasants, and not merchants, for perhaps even thousands of years. We were told that besides Sandor there was only one other Jewish village of this type in Kurdistan.

Soon afterwards we had an even stranger experience. Riding out from Saratuk we also visited several villages of Chaldean Christians who spoke an Aramaic dialect which was even closer to the language of the Talmud than the Assyrian vernacular. In these and other nearby villages, we came across Jewish families who were serfs. They were extremely poor as they were not allowed to own cattle or land, and could not move around freely. These people also expressed a desire to come to Palestine as soon as possible and this wish was eventually realized.

We then went through the Rawanduz gorge to Rayat, where the magnificent waterfalls reminded me of the valley of Lauterbrunnen in Switzerland. The mountains of Rayat, which reach up to 4000 m in height, are on the Persian border. The entire area is covered by dense forests, which we studied. The inhabitants were semi-nomads. At full moon, we reached the sheikh's tent, which had a beautiful view of the mountains covered by snow. The sheikh treated us to a large meal consisting of roast mutton, rice, vegetables, yoghurt, buttermilk and pittah (flat bread) served in large bowls.

The next day, in an attempt to climb one of the mountains, we crossed the Persian border by mistake. We quickly retraced our steps when we saw Persian soldiers running after us. We got back in time to avoid arrest by the Persians, who were on bad terms with the Kurds at that time.

The following day we went back to Jerusalem.

Chapter 4
Life in Palestine and
Military Service during the Second World War

In Palestine in addition to my work for the Aaronsohns, I intensified my studies in desert ecology (24), focusing on the effects of soil and habitat on the distribution and anatomical structure of the plants growing in such environments. I encouraged my students, Alexandra Poljakoff, E. Shmueli and Ephraim Konis, to begin investigations on water balance in olive trees and plants in the salines near the Dead Sea, and on the heat balance in various other local plants. The Department's Physiology Section developed into a center for Ecophysiology of the entire Middle East.

In addition, I was confronted by aspects of plant physiology, which would not have captured my imagination in any other place in the world. Following the destruction of the Second Temple in Jerusalem by the Romans in 70 A.D., the Jews, who had developed flourishing agriculture in Palestine, uprooted from their land, were dispersed all over the world. In the Diaspora, they were not allowed to own land, and so through the force of circumstances, were cut off from farming for 2,000 years. Consequently, when the first Jewish settlers came to Palestine at the end of the nineteenth century, they had by then lost all agricultural tradition, which had both advantages and disadvantages. On the one hand they were not bound to antiquated agricultural practices handed down over the generations while, on the other hand, were completely lacking in experience. Many of these "new peasants" were University-trained and therefore attempted to work the land in a scientific manner. In this regard, David Zirkin, a member of Kibbutz Ayn Harod turned to me for advice in developing a scientific method for rooting cuttings. His inquiry led to long and fruitful cooperation between him, my assistant Konis, and myself, in which we put Molisch's "teachings", relating to the importance of applying the principles of Plant Physiology to gardening and agriculture, as propagated in his well-known book "Plant Physiology as Theory of Gardening" (25) into practice.

We were the first group in the Middle East to use plant hormones for rooting of cuttings and for achieving grafts of scion to rootstock. We found new ways of bringing about early flowering in

Irises prolonging flowering time. We developed a method for stimulating Gladiolus bulbs to sprout. Our research group also applied cold treatment (vernalization) and other physiological techniques in order to help the farmers to introduce the potato crop into this country.

Through my work with Zirkin, I also went back to my original fields of interest. He had problems in germinating seeds of various agriculturally important plum and apple varieties, which we managed to resolve. This led me back to my old love: germination. Thus germination and germination inhibiting substances became the major research topic of the Physiology Section for many years to come.

Two early papers in this field were the work of Gershon Mosheoff one of my students. He discovered a substance, which can be extracted from wheat kernels with water, which first inhibits, and then stimulates germination. Regrettably these were both the first and last studies of this promising young man, who was killed in an Arab attack on Kibbutz Kiryath Anavim in 1936.

Most of these papers in Applied Plant Physiology were either published in the agricultural journal "*Hassadeh*" ("The Field"), or in the Department's "Palestine Journal of Botany, Jerusalem Series," established earlier by Oppenheimer as "Rehovoth series" and financed out of his own pocket. For many years, every Department member "donated" a small sum from his meagre salary in order to support the "Jerusalem series." These two publications were later combined to form the "Israel Journal of Botany," which is still active today.

These projects in Applied Plant Physiology gave me the feeling that I was making some contribution to the physical development of my new homeland. As a University scientist, I appreciated the rare privilege of working in a field which interested me both intellectually and emotionally. This freedom, to choose one's field of work gives us a duty to repay society by directing research toward resolution of social and other problems of society, thereby creating a better future for mankind.

In spite of all the Zionist "schooling" I had received, Palestine was a strange land to me when I came to live here in 1933. Although I could not say that I had come home, I soon came to feel that in time, Palestine would become my country. I was an equal among equals, a proud Jew among Jews, who worked in every conceivable field, farmers, builders, factory workers, policemen, drivers, waiters, and fishermen. We were idealists, eager to build our own country.

Daily life in Palestine was tough at that time. The land had been neglected for hundreds of years, the original forests destroyed, and Jerusalem was surrounded by stony, depressing wasteland. Water supplies were scarce and rationed. Returning from the desert, tired and dirty, I often did not have the luxury of cleaning myself because of lack of water. Each drop of water was precious, and had to be used more than once, if possible - first for personal hygiene, then for cleaning the stone floors, and finally for flushing the toilet. We were often without electricity, and we cooked on small "primus" stoves fuelled by kerosine. For heating a primitive oil heater was all we had to take the edge off the cold in winter.

We soon learnt to cope with these physical discomforts, but found the consequences of the Arab-Jewish conflict more difficult to face. The British police were both unwilling and incapable of protecting us against Arab attacks. At the time we lived in Kiryat Shmuel, which was a small isolated suburb on the outskirts of the city. One night I met Adolf Reifenberg, an Assistant I knew from the University's Biochemical Institute. He told me that he belonged to the *Hagannah* (the Jewish self-defence organization). Soon after, in the fall of 1933, I joined the *Hagannah*. On that particular day, Arabs had attacked and killed several Jews in the Old City of Jerusalem and there was fear of an all-out Arab attack. I served with the *Hagannah* until 1948, when it was malgamated with other fighting units to form the Israel Defence Forces (IDF).

I had to learn how to use guns, which we kept carefully hidden from the British police since possession of fire arms was an offence. I did guard duty two to three times a week, and found it difficult to lecture at the University after sleepless nights. In time, I was made responsible for the security of one of the outlying suburbs, which was an additional heavy physical and emotional burden.

However to me, "loss" of German, one of my mother tongues, was the greatest handicap. Only people who have gone through similar experiences can understand what this really means - I was robbed of the language in which I thought and in which I published my scientific work. I quickly had to learn to lecture and speak Hebrew, and to write in both English and Hebrew.

I had not learnt English in school and had to teach myself the language. Although I had gained a rudimentary knowledge of Hebrew in the Blau-Weiss Youth Movement, this was a far cry from being able to present a lecture in Hebrew. After a few Hebrew lessons from my assistant Shmuel Duvdevani, I delivered my first lecture in Hebrew with many mistakes, which were a source of great amusement to my audience.

There were no textbooks in Botany available in Hebrew at that time. Only much later, together with my Assistant Konis, I published two textbooks in Hebrew on Plant Physiology in stencilled form, and a short printed book on General Botany, and Michael Zohary, Konis and I translated the book of Timiriasew entitled *The Life of Plants* from Russian into Hebrew.

In view of my personal experiences in Darmstadt and the treatment of Jews in Germany, I decided to get rid of my German name Walter Schwarz. When I became a Palestinian citizen in 1935, I gave myself a Hebrew name: I chose "Michael," the name of one of the archangels, as a Christian name. *"Shahor,"* the Hebrew for my family name, "Schwarz," did not appeal to me. When I told the official this, he asked me my mother's maiden name, and informed me the Hebrew equivalent for "Loewenstein" would be *Evenari*. I liked the sound of this name, and the idea of "preserving" something of the interesting history of this part of my family. The decision to Hebraize my name was not an easy one since Walter Schwarz had already published 17 scientific papers, and was mentioned in Möbius' History of Botany (6), whereas Michael Evenari was a complete newcomer to Botany!

Two years after changing my name, Ester (nee Gabriel) gave birth to our son Eliyahu, whom we called Eli for short. He had a difficult childhood as Esther and I separated when he was very young. Eli always showed a great love for animals, and I am sure would have made a good veterinarian. However because of a ridiculous quarrel with his teacher of religion, he was barred from taking matriculation examinations, and therefore could not go on to study at University. After completing his three years of compulsory army service he could no longer face taking examinations. Since he was gifted artistically, and liked to work with glass, he went to a vocational school in Germany, which he completed with very good marks.

When he came back to Israel, Eli met a young German girl named Christel Eckern. She had previously visited Israel with a group called "Aktion Sühnezeichen" (Action Repentance), an organization that aimed at making amends for the way the Germans had treated the Jews. On her return to Germany, she wrote an interesting book called "The Road to Jerusalem" and sent a copy of it to the Israeli Ministry of the Interior with a request to become a temporary resident, which was granted. She returned and worked in Nazareth in a home for orphaned Arab children, established by the Greek Orthodox bishop. During this period, she visited Eli's mother Esther, a leading social worker at that time, in order to ask for some professional advice. She met Eli there, and they fell in love and got married. They now have two sons, Jotham and Joel. Regrettably, they found it difficult to make out here in Israel and they now live in Bavaria.

Apart from the trip to Kurdistan, the Department of Botany made several other excursions in the Middle East in this period. Two of these were particularly important for my future.

In 1936, on the way to Aquaba, we visited the former capital of the Nabateans, Petra. The tombs and monuments hewn into the multi-colored Nubian sandstone, one of the Seven Wonders of the World, were a breathtaking sight. This tourist attraction today is easily accessible but at that time it was difficult to reach. The numerous water works, channels and cisterns there caught my imagination. Who were the Nabateans? What was the water source in this desert city? Little did I realize that the culture, history, and source of water supply of this long-forgotten people would later become one of my main scientific interests.

On the way back from Aquaba, we went through the rough terrain with very few roads of the Negev desert. I never imagined that I would one day come and live in that area.

The other important excursion was into the Sinai Desert in 1940. On the way back from the Monastery of St. Catherine in Sinai we met some Egyptians, who told us that the "phony war" had come to end with the German invasion of Belgium and Holland. At the border with Palestine, Gurkha guards arrested our party, claiming that we were German spies. It took us two days to convince them that we were *bona fide* botanists.

One of the secondary aims of the Sinai trip was to collect seeds of *Hyoscyamus muticus*, a poisonous species of henbane, which contains medically important alkaloids. Palestine had been commercially isolated from Europe since the outbreak of the Second World War and the government was therefore interested in producing priority items such as war material and drugs locally. The plant grew well in a Kibbutz. Unfortunately, some of the cows started grazing in the field of henbane and died, and for quite some time, I was ashamed to show my face there. An attempt to grow a species of Valerian of the upper Galilee, which we collected there, proved a good substitute for the European Valerian. A Kibbutz cultivated it successfully and a pharmaceutical factory produced from it the sedative antispasmodic essence which is used in medicine.

Shortly after our return from Sinai, I volunteered to serve in the British Army, which had very few troops in Palestine at that time. I joined the "Palestine Light Anti-Aircraft Battery" (Pal. Light A.A.Bty for short), a unit of Palestinian Jews, all members of the Haganah, mostly from *Kibbutzim*.

As Jews, we felt a strong desire to show Hitler that we could fight back. In addition, Rommel's approach to Egypt posed a potential threat of annihilation of the Jewish population in Palestine. Together with the Haganah, we were ready to form a guerilla unit in preparation for facing the Nazis if necessary. Later, our unit also fulfilled roles in Cyprus and Europe. During the war, an unofficial armistice prevailed between the Jews, Arabs and the British in Palestine. Even so, the British did not allow Jewish refugees into the country. Survivors of concentration camps and Jews who had survived the war by going into hiding were interned in camps by the British, if caught trying to get into the country illegally. We did everything we could to smuggle these people into Palestine, which under the circumstances, I felt was far more important than my scientific research. Sometimes, we would dress them in a British uniform and gave them false papers so that they appeared to be soldiers on leave.

We also smuggled in any arms that we could lay our hands on, some of which came from Italian partisans. These later proved to be crucial for the survival of the Jews in Palestine. In the face of the British pronouncement that they would not allow substantial immigration of Jews into Palestine even after the war, we felt certain that the conflict in the region would flare up again. It seemed to us, that we would then confront a bitter all-out struggle, and therefore did everything we could to prepare ourselves for that eventuality.

I enlisted at Sarafand, which served as the British Army recruitment center at that time. I was given a uniform and, without further preparation, was immediately assigned to a unit stationed in Haifa. Upon my arrival there, I immediately joined the Anti-Aircraft (A.A.) unit, not even knowing how to salute properly, I found myself in active service.

The British came into the war unprepared. Since they did not have any A.A. guns in Palestine, we had to make do with small "pea-shooters," small guns found on an Italian ship captured while carrying arms to Mussolini's armies in Lybia, and to use our own initiative to find out how they worked. In this respect, Hans Jonas, who is now a well-known Professor of Philosophy in the United States proved to be remarkably inventive, and even devised a sight device for our "pea-shooters." It was only much later that we got the standard Bofors A.A. guns.

Several times we went into action with our "pea-shooters" providing the only form of defence against Italian, German and Vichy aircrafts stationed in Lebanon, attacking Haifa. We were installed in detachments of six on the roof of the Haifa railway station, on the breakwaters in the harbor, and around the oil refinery near Haifa, and also at the Potash factory by the Dead Sea, which was a potential target of enemy bombs.

Life in the army was quite good, since according to British custom, each detachment had its own cook and kitchen. As compared to the German "Gulash-Kanone," which served a whole battery, we had good meals prepared from the daily rations we received. My specialty was baking cakes, which were greatly appreciated by the rest of my detachment.

During one of the air attacks on the harbor, a bomb exploded in the water bringing hundreds of fish to the surface, and just missing our gun position at the breakwater. We collected the fish, baked as many as we could eat and exchanged the rest for other food items at the local market.

Our battery, consisting of students, former students, Kibbutzniks or future generals in the Israeli army and future Professors (Jonas, myself and some others), was a strange artillery unit indeed! Most of our British N.C.O.'s, with whom we had excellent relations, were drafted workmen. Some of our men, such as Avraham Joffe, who later became one of the most successful generals in the Israeli army, understood more about military matters than the British officers. As far as I can remember, one of the "King's Regulations" stated that only British subjects could become officers in the "Royal Artillery," whereas in the Palestinian-Jewish infantry units, Jews did serve in that capacity.

The British officers did not always know how to handle our bizarre unit. Who had ever heard of a battery that brought out a monthly "newspaper" in Hebrew containing articles of scientific interest, among others? In one of these I described how one could, observing the stars, during long night watches, determine the time to the nearest five minutes.

None of our privates were prepared to serve as a "bat man servant" to an officer - a position that was much sought after in regular units. On one occasion a new officer asked me as the Battery Sergeant

Major (B.S.M. for short) to provide him with a bat man. At the next morning parade, speaking in Hebrew, I asked for a volunteer since according to the regulations one could not order someone to be a "bat man." When nobody came forward, the officer explained his request in English. Still no one moved, and he threatened bringing a British private to do the job. Since we wanted to remain a purely Jewish unit, our unofficial Soldiers Council met that evening, and decided that one of the *Kibbutzniks* should "volunteer." A day later, the officer stormed into my tent, and foaming with rage, said that he was going to bring his "bat man" to court martial. Apparently the new "bat man" refused to take a tip from the officer after cleaning his uniform and polishing his buttons with "Brasso." The *kibbutznik* considered this "baksheesh" to be an insult and threw the money at the officer. How times have changed! Today taking tips is quite normal in Israel. With difficulty, I succeeded in explaining to the officer, that a "tip" was an offence to a *kibbutznik*. All in all, our relations with our officers were very good, and some of them became real friends and I still correspond with them. Off duty, officers and men alike, referred to Jonas and I as the "Profs."

Once, when I was on night duty guarding the headquarters, our major, an Irishman, came back from town stone drunk. After the customary identification procedure tapping me on the shoulder, he said: "Prof, after the war I'll come back to Palestine and help you to get rid of the damned British!"

After a stint in Haifa, our battery was transferred to Cyprus. In the meantime the Germans had conquered Greece and Crete, with the exception of one island of the Dodecanese, and the British thought they would also occupy Cyprus. They could have done so easily since it was only defended by our unit and a few other British troops. However, strangely enough most of the time they left us alone and from time to time, German planes would cruise over the island, usually out of the reach or our Bofors. We were also responsible for protecting a tiny Dodecanese island very near the Turkish coast, which was code-named "Trombone." Since it was so near to the Dodecanese occupied by the Germans, we could only go there at night in small boats, and we had to be very careful not to step on the many mines in the dark.

Our main role in Cyprus was to protect the airport near Nicosia and the port of Famagusta. Having been promoted to the rank of Sergeant, I was now in charge of one of the guns. My detachment was positioned at one of the most romantic points, the old Venetian walls near the "Othello Tower," which still carries the winged Lion of St. Marcus on its portal. In the old city, nearby, there were the remains of the church of St. George, and the beautiful Gothic Cathedral of St. Nicholas, on to which the Turks had built a minaret. Naturally, these surroundings aroused my archeological and historical curiosity, a condition which sooner or later affects everyone living in Israel, and soon I had infected the rest of the battery with this "disease."

My best friend Ebs (short for Ebstein) and I visited the small, well-organized archaeological museum in Nicosia, and met the Director, Dr. Mugabgab, with whom we became firm friends. He often invited us to his home, and through him we got to know about the fascinating and caleidoscopic history of the island.

As early as 6000 B.C.E., the inhabitants of Cyprus, whose origin is unknown, practiced agriculture. Copper was mined and exported from there in chalcolithic times. Later, the island was successively colonized by Mycaeneans and Achaeans, who, according to some views, came from Troy, and later Dorean Greeks. The Cypriot Greeks of today originated from the latter, although the Phoenicians, Assyrians, Egyptians and Persians all subsequently ruled the island for short periods, until it was conquered first by the Ptolomaeans and then by the Romans.

Cyprus has several historical links with the Jews and Christians. Under Roman rule, over 200,000 Jews lived in Cyprus, most of them in the city of Salamis, the main town and port on the Island at that time. The Jews of Cyprus revolted against the Romans in 116 A.D. - a foretaste of the last stand of the Jews of Judea under the leadership of Bar-Kochba (132-135 A.D.). Thousands of Jews and Greeks in Salamis were massacred in this bloody revolt. Furthermore it posed such a serious threat to the Romans that they expelled all the Jews from the island and published a decree forbidding them to enter the island under any circumstances including shipwreck. Seeing the ruins of Salamis for the first time evoked a similar feeling in me as the Arc of Titus in Rome, which was meant to immortalize the Roman triumph over Judea and Jerusalem. How ironic it is that whereas the Roman nation had vanished without trace, the descendants of the people they had sought to destroy are still around!

Jewish and Cypriot histories also crossed at another point as I learnt from a visit to the Monastery of St. Barnabas, the Patron Saint of Cyprus. Born in Salamis, Barnabas was originally a Jew. After coming to Jerusalem, he became a follower of Jesus and on his return to Cyprus, together with Paul, he converted the Roman Pro-Consul to Christianity. Thus, two Jews brought the teaching of another Jew, Jesus of Nazareth, to the island.

During 800 years of Byzantine rule, the island flourished. Many of the monasteries and basilicas with their magnificent frescoes, mosaics and icons date back to this time.

During the Crusades, Cyprus went through the most romantic period in its history, and owes the pure beauty of its Gothic buildings, such as the Cathedral of St. Nicholas, to that time. Another fascinating sight is the almost perfectly preserved Knight's Castle at St. Hilarion, standing on a plateau in the jagged Kyrenia mountains surrounded by forests. None of the Medieval castles I have seen, either in Europe or Palestine, compare with this.

During the third crusade Richard the Lion Heart conquered the island. After the Templars, who were interested in buying the island, did not come up with the money, Richard gave it to Guy de Lusignan, the French Crusader, who was destined to be King of Jerusalem (although he never realized this position, his successors retained the title "King of Jerusalem"). This dynasty imposed a feudal rule in Cyprus, and was responsible for erecting all the beautiful Gothic buildings there.

The last king of Cyprus, James de Lusignan, married Caterina Cornaro, the daughter of a Venetian nobleman. She became Queen of Cyprus after his death. In 1489 she abdicated, giving over rule of Cyprus to Venice, and received a Venetian castle in recognition of this patriotic act. As a former Queen, when she died she was buried in the Church of St. Marc. When we were in Italy I visited her grave for sentimental reasons.

The Venetian rule of Cyprus was brought to an end by the Turkish conquest of the island in 1571. After a long siege, they occupied Famagusta and massacred all the soldiers there although they had promised not to harm them. They laid waste the old city, which has since remained in ruins.

Under Ottoman rule, many Turks immigrated to Cyprus, who brought Islam to the island. Even after 400 years of uneasy co-existence, these two nations still hate each other as reflected in the present state of affairs. As in Palestine during their rule the Turks neglected the island. Cyprus remained in Ottoman hands until 1914, when the British took it over as a crown colony.

Through Mugabgab's vivid accounts, background readings and organized trips either in groups or just with Ebs, we relived the history of the island. As a Sergeant-Major, I had a motorcycle, and with Ebs on the back, we travelled the length and breadth of the island, and where there were no roads, we walked. I always looked out for plants, focusing on the many endemic plants. We found the most interesting flora on the igneous alpine rocks in the Troodos mountains and the extended chain of limestone ridge stretching from the magnificent monastery of Bella Pais to Rhizokarpas at the north-eastern tip of the island. The herbarium I collected I sent to the Hebrew University, where it can still be found today.

In what was in my view the most beautiful excursion of all, we walked from the summit of the Olympus mountain by way of the Kykkos Monastery to the Bay of Morphou, and then on to Kyrenia, a jewel on the Mediterranean coast, where many retired British officers and colonial staff lived in luxurious villas surrounded by nice English-style gardens. For hours on end, we trekked through the dense forests of Cypriot cedars, which I found much more impressive than those I had seen in Lebanon. According to the Director of Forests in Cyprus, who helped me with my botanical work, this forest contained about 40,000 trees.

We came into close contact with both Greeks and Turks, and felt the profound tension and enmity prevailing between them. The Turks were unfriendly, while the Greeks were more open, especially when they realized that I could read and speak their language quite well. I soon discovered that if I quoted verses from the Odyssey, the Iliad or the plays of Aeschylus, Euripides or Sophocles at the Greek taverns, I was treated to good cognac ("Aphrodite"), wine ("Commanderia") or Arrak ("Uzo").

My initiation into riding a motorcycle, which I used so much in Cyprus, and later in Italy, was somewhat similar to my horse "riding lesson" in Kurdistan. As a Sergeant-Major, I was supposed to ride a motorcycle, and duly received a heavy "monster" of a cycle, a Harley-Davidson, if my memory serves me correctly, being ashamed to reveal to my superiors that I had never ridden a motorcycle. I asked my colleague Sergeant-Major Haas to show me in a few hours how to use it, how to lift it when it fell down, how to sit on it, how to give gas and how to brake it. I then sat on it and roared off at full speed. Miraculously, I survived, although I never felt completely at ease with the "monster", which went everywhere with me until the end of the war.

While in Cyprus, we often visited a camp in the Troodos mountains in which "illegal" refugees captured by the British on their approach to the shore of Palestine were interned. According to the infamous white paper Jewish immigration into Palestine was strictly forbidden. We tried to help them as best as we could, smuggling many of the younger ones into Palestine. One of the internees, who impressed me with his remarkable intellectual faculties, is now a Professor at Tel-Aviv University.

Since the beginning of the Second World War, politicians had attempted to convince the British government to incorporate all the Jewish Palestinian units in the Royal Artillery, the "Buffs" (Infantry), and the Royal Engineers, among others - into a unified Jewish Brigade which would fight Hitler under the blue and white flag bearing the Star of David. Although Churchill supported such a move, most of his Ministers, fearing a negative reaction from the Arabs, opposed him in this issue. Eventually the Brigade was formed in Burg-el-Arab, near el-Alamein in Egypt's Western desert. Our unit left Cyprus for El-Alamein to join the new Jewish Brigade. From there we were sent to Italy, and in transit on the ships we manned anti-aircraft guns for the last time.

Chapter 5
With the Jewish Brigade in Italy

In Italy, our unit was incorporated into the Field Artillery Regiment within the newly formed Jewish Brigade. About half the men were Jewish/Palestinians and the rest British. The British were furious at being separated from their comrades, with whom they had fought throughout the war, and attached to such a strange "Jewish" unit. In spite of being a Jewish Palestinian and not British, I was in charge of British troops. This as far as I know was strictly against the King's regulation which stated that only British subjects can be in charge of British soldiers. Being the most senior B.S.M. I was placed in charge of British soldiers and made head of the Sergeant's mess. Although I sympathized with the British soldiers it did not help matters.

From the first meeting in the Sergeants' Mess Hall, these feelings emerged clearly. The working-class British sergeants were already completely drunk and when we came in, one of them, a bricklayer in civilian life, declared *coram publico:* "You are all f..king Jews, all f..king Jews are f..king cowards." This hit one of my most sensitive points, and had he not been stone drunk at the time, I would have certainly struck him. The next morning I confronted him: "Now that you are sober, would you please repeat what you said last night about the Jews?" Everybody stopped talking and looked at us, while he remained silent. "So," I said, "then *you* are a coward who does not take responsibility for his own words." This cleared the atmosphere, showing the British that we would not stand for anti-semitic remarks. In time, our relationships with our British "colleagues" improved. However, the bricklayer never forgot our confrontation. He tried to take revenge several times, which could have cost me my life. In contrast, as in our A.A. battery, we had good relations with the British officers in the Brigade.

In my daily contacts with British conscripts, I compared their behavior to that of our officers and of the American enlisted men I came across. This opened my eyes to the complete separation of the classes among the British - the "upper" and the "lower" classes. They speak distinctive languages and have different norms of conduct. In particular the privates seemed to show a complete lack of ambition to better their social condition. When I asked what professions they would like their sons to have, they would invariably answer: "Naturally, the same as I am." During my career in the British army, I only came across one officer who had been promoted from the ranks. How different this was to the attitude among the American conscripts and working men, whom I met later in the United States! How much has the situation in England changed since that time?

I also learned there was a considerable difference between British conscripts and professional soldiers. We underwent retraining to learn how to use 25-pound field artillery guns, mastering the details very rapidly. During this period, I shared a tent with two other Sergeant-Majors from the professional army. They had served for years in India, Hong Kong, Kenya, Nyasaland and many other exotic places. Very interesting people, self-educated, with a broad outlook on life, they told me many fascinating stories about their experiences. We became good friends, and I cherish their memory.

Whenever we were not in action, I took the opportunity to collect plants. I got hold of an analytical key to the flora of Italy in a bookshop in Naples and also sent this collection to the Hebrew University.

During this period, I spent some time in a Roman military hospital with a broken kneecap. One day, when I had a chance to get out for a few hours, I decided to visit the Vatican at a time when by chance Pope Pius XII happened to be giving an audience to soldiers from the allied forces. He gave a short speech, and then walked along the first row of the audience where I, heavily bandaged, had been given a place. As he passed by, people knelt down and kissed the fisherman's ring, but as non-Catholics, two of us, a British officer standing next to me and myself, remained standing. The Pope stopped and asked me in English: "Where are you from, my son?" On the spur of the moment, wanting to show off my Latin I answered: "Terrae sanctae civis sum Judaeus. Tibi gratias ago nomine populi Judaei quia salvabas vitam Judaeorum tam multorum" (I am a Jew from the Holy Land. In the name of the Jewish people, I thank you for saving the lives of so many Jews). Although after the war, the Pope was sharply attacked for his passive attitude towards Hitler's regime, we came across many Jewish families, who had been sheltered in monasteries and saved from the Germans at the Pope's order. Slightly taken aback, the Pope looked at me and asked in Hebrew: "Then you speak Hebrew. Let me bless you." He spread his hands out over my head in the manner of the priests of the Jerusalem Temple and then recited the priestly blessing in Hebrew: "The Lord bless thee and keep thee. The Lord make His face shine upon thee and be gracious unto thee. The Lord lift up His countenance upon thee and give thee peace" (Numbers 6:24-26). I will never forget his ascetic spiritual face and the beauty of his slender hands. Other members of my unit, who were also present in the audience, later spread around the following bon-mot: "Who was that priest who talked to our Evenari for so long?"

On Armistice Day in May 1945, our unit was near Palmanova, a small town south of Udine in northern Italy. Naturally we celebrated the great event, the Jewish and British members holding separate parties. The Jews held a party with singing, dancing and drinking in a forest. Suddenly, one of our guards on the look-out for German soldiers who had not yet surrendered, came running up to us, shouting: "The British are burning our flag!" This was like waving a red rag in front of a bull. It had taken years to get permission from the British to form a Jewish Brigade with its own blue and white national flag bearing the yellow Star of David, an emblem the Nazis forced the Jews to wear as a sign of shame. Today it may be difficult to understand the pride we took in our flag, later adopted by the State of Israel, which gave hope to the survivors of the Holocaust. In a frenzy, the men took up their guns with the intention of attacking the British party. Fortunately one of the sergeants, several British officers and I succeeded in calming our men down. If we had not succeeded there would have been a massacre. An investigation by the officers revealed that "my" bricklayer had instigated this action. He was court-martialled and severely punished. This was the last contact we had with our British "comrades," since only the British officers stayed on with the unit.

After the Armistice, we had little to do. We asked the commander to allow us to prepare for our return to civilian life. Most of the soldiers in our unit were farmers or students of Agriculture, so we wanted to do some agricultural work. I was asked to organize this venture. We went to Fagagna, a small, romantic village in the Italian Alps, close to the Austrian border, where we lodged with the peasants. In the mornings, the "students" worked in the fields, and in the afternoons, I gave a course on the Application of Botany and Plant Physiology to Agriculture. In my free time, I collected plants in the mountains. Most interesting was the trip to the Triglav on the Italian-Austrian-Yugoslav border, where I had the incredible experience of seeing a whole field of Edelweiss, usually a very rare alpine plant at a height of about 2800 meters. Apparently, because of the war, people had not been up into the mountains, and the Edelweiss was able to grow unhampered by man.

In another interesting excursion, Ebs and I went to Mallnitz. We climbed the "Hochalmspitze" (3360 m) in the Austrian Alps. Before ascending the summit we spent the night in the "Hannoverhaus," an alpine hut for mountaineers. To our amazement, we found it full of smartly dressed, well-fed men and women. Slowly, it dawned on us that these people must be high-ranking Nazi officials hiding in the mountains in order to evade the Allies. Upon seeing our British uniforms and our shoulder epaulettes bearing the Star-of-David emblem of the Jewish Brigade, they became visibly frightened. Among themselves, they discussed what to do with us, not knowing that we understood German. We felt scared and slept in shifts behind a door which we wedged up with table and chairs, guarding ourselves with loaded pistols. On our return to the unit, we reported the whereabouts of these Nazis.

Ebs and I also went on an excursion to the "Grossglockner" glacier which I had seen before. It was difficult to reach the mountain, which is some distance from Fagagna. I had to stop German military lorries driven by unarmed German soldiers for lifts along the route. Standing in the middle of the road, with one hand pointing at my B.S.M. badge which was a leather bracelet bearing a golden crown surrounded by laurel leaves, and the other at my epaulettes with the emblem of the Jewish Brigade, which was well-known to German soldiers, I would shout "Halt!" in German. The frightened drivers would stop immediately, and I would then get in at the front next to the driver and Ebs would take a back seat. We would then order him to drive to where we wanted to go. None of the drivers we came across dared to refuse our requests. A tall soldier from New Zealand once observed this procedure, which was, of course, highly illegal. "Good for you, boys!" he said, saluting us. By this means, we finally arrived at the Grossglockner.

After the Armistice, we started searching for relatives whom we hoped had survived the Holocaust. These were sometimes daring and dangerous operations since we often had to go into Russian-occupied zones. Although our British officers unofficially knew what we were doing, they never interfered with our efforts. We will never forget their "silent" help in this matter. The story of Pauncz, a soldier in our unit, serves as a good example of these activities. He was a Hungarian Jew and wanted to go to Budapest to search for his parents. As the first step, I had to provide him with false leave passes. Pauncz then removed his epaulettes so that he would look like a "regular" British soldier, and hitched to Frankfurt. From there, he took a German train to Würzburg, traveling in a compartment reserved for American soldiers. The train stopped in Pilsen, Czechoslovakia, bordering on the Russian zone. Pauncz got off the train and fortunately met some Jewish soldiers of Czech origin, who had volunteered in Palestine to serve in the Czech army in order to return to their "homeland," but after they had seen what had happened in Czechoslovakia, they wanted to return to Palestine. They helped Pauncz board a train to Budapest. This time he was in a compartment with Russian soldiers, who took him for a British officer on an official mission. Near Budapest, the train stopped and everybody had to get out to be checked by Russian military police. As Pauncz stood in an open field surrounded by Russians, a Hungarian Jewess carrying a child approached him, first addressing him in Hungarian, and then in Yiddish. Since Pauncz was supposed to be British, he did not dare to answer her. However, suspecting that he was Jewish, she said to her child: "He understands me, but doesn't want to say that he is Jewish." After a long palaver, the Russians allowed him to reboard the train. Eventually he arrived in Budapest, where he found his mother, who fainted when she saw him at the door. He found out that his father was dead, and that all his other relatives and friends had vanished during the Holocaust. He was deeply shocked by the anti-semitic attitude among the Hungarians, many of whom had helped the Nazis to deport Jews. After doing everything he could to ensure that his mother could leave for Palestine, he returned to the unit.

My own search revealed that thirteen members of my family had been deported from Germany, Holland and France, and gassed in Auschwitz. Their non-Jewish neighbours told me all the details of how the SS-men had taken them from their homes. Only six of my closest relatives had survived in France by posing as peasants in remote villages in which the residents protected them from the Nazis.

I will not dwell on what I saw in the concentration camps, since I find it too painful. However, I will present a few stories, which I noted down, related by Jews with whom our paths crossed briefly. The specific "vocabulary" of former concentration camp inmates shocked me when they referred to "Kapos", "slaves", "muselmen" (inmates who were close to death) and "prominents" (inmates favored by the SS) as if they were common words in the German language.

Man with the Violin

On the return trip from the Hochalmspitze, we were delayed in Spittal on a Saturday. We met some British soldiers who recognized us as fellow Jews from the Brigade emblem. They were on their way back to Villach from a Sabbath service for Jewish British soldiers and offered us a lift. They were also carrying two civilians. The younger, who was around 40 years old, was wearing some items of an American army uniform and holding a violin case tightly under his left arm. We spoke to them in German, asking them all the usual questions: "Are you Jews? Where do you come from? Where do you want to go?" The man with the violin answered with a flood of words as if there was no time to lose. He was very nervous and his face constantly changed expression. A married man with two daughters, he had been a well-known professional violinist from Prague, played in the best jazz bands and earned well. The Nazis first deported the whole family to Theresienstadt, where he was separated from his wife and children, whom he saw last on their way to their death in Auschwitz. He could not give them a sign of recognition. In order to save his own skin he pretended to be 28 years old and single. First he worked as a "slave" in a factory. When it was discovered that he played the violin, he was forced to perform at parties held by the SS-men: "Jud," they would say "play one of your nigger jazz dances we like although they are banned by the Führer." Although an assimilated Jew with very little Jewish background, he then remarked: "You know, when I played for those beasts, I vaguely remembered something written in the Bible about Jews being forced to play to their enemies, which was a source of consolation to me. On leafing through a pocket version of the Book of Psalms I found the following words: "For there they who carried us away captive, required of us a song, and they who wasted us required of us mirth, saying: 'Sing us one of the songs of Zion!'" (Psalm 137). He was later transferred to Dachau which was liberated by the Americans before the Nazis could carry out their plans to kill the surviving inmates. Since he spoke English well, he helped the Americans to search for SS-men, who had fled into the surrounding countryside. His entire body shook with deep hatred as he described catching some of these Nazis and killing them.

In return for his help, an American officer had given him the real Stradivarius violin that he clung to so fervently. The officer had "liberated" it in Munich. We asked him what he was going to do now:

"First I have to get rid of the 'barbed-wire feeling' by traveling around. Then? I have no idea!" The three of us got out of the car at Villach. Sitting down by the roadside we ate bread and bully-beef and drank some wine, and suggested that he make his future in Palestine. He showed no interest in this possibility. With an absent look on his face, he shook hands with us as we parted - he took a lift to Klagenfurt and we went on to Tarvisio.

The "Aryan"

The older man, who was in the car with us on the way to Villach, was a rich Jew from Dahlem, one of the wealthy suburbs of Berlin. His neighbor, a high-ranking police officer, arranged false "Aryan" papers for him and his wife in return for 70,000.- DM. As an "Aryan," he described dispassionately how he looked on helplessly as his sister and her family were deported from the next house. He told us this without visible emotion. This unnatural apathy, which we also observed in others, was apparently a cover-up for a deep feeling of guilt. As more and more Jews were deported from Berlin, on the advice of the police officer, this Jew had left the city for Villach, where he was still living under his assumed name. Why had this "Aryan" attended a Shabbath service? He had no answer. "I feel as if I had climbed through a barred window into a garden of freedom, having got there, freedom has lost all its meaning for me. Maybe, the price I paid was too high."

The Suspect

Once, while waiting for a lift in the shade of large Linden trees, in the Austrian village of Möllbrücke, we spotted an athletic looking young man on the other side of the road. Noticing our Brigade emblem, he came over to speak to us. A German Jew from Halberstadt, he had succeeded in getting hold of Aryan papers and a passport, and was living in Möllbrücke with his non-Jewish girlfriend. How had he in the guise of a young healthy Aryan, evaded serving in the German army? He claimed that he managed this by keeping on the move. His story seemed difficult to believe and I suspected that he may have been an informer, who denounced those in hiding to the Gestapo. He flooded us with questions: What should he do now? Should he confess to being Jewish? Where should he go? We promised to give his address to various people who might be able to help him and we kept our promise. When a car travelling in our direction picked us up we left him standing in the street. What may have become of him?

La Bionda

In Palmanova we met a Jewess from Graz in Austria, who had married a non-Jewish Italian living in Trieste. With her golden blond hair and Aryan features, she pretended to be Aryan. When the Germans occupied Italy, she had separated from her husband and left Trieste for Palmanova. The Germans turned the barracks in Palmanova, which we later took over for a short period, into SS headquarters and "Security Service." The large courtyard behind the building was surrounded by a high wall full of bullet holes - silent witnesses to the many victims who met their deaths there. Bionda got herself a position as a German-Italian translator for the Germans.

At the beginning, the Germans didn't trust her completely. But after she had won their confidence, she handled many secret documents through which she found out in advance who the Germans planned to arrest. She managed to warn people and saved many lives. First we did not believe her story. Later several Italian partisans from the surrounding area who shared the barracks with us for some time, confirmed her story and told us how they owed their lives to her. She was eventually betrayed by an Italian, who was later shot by the partisans but she managed to escape over the mountains into Switzerland. After the Armistice, she returned to Palmanova to look for her furniture and belongings, which the Germans had confiscated. We helped her, and found most of her property, which we then transported to her home in Trieste. We also went to Graz, to find out what had happened to the rest of her family - all of them had been deported and she was the sole survivor.

Modern Marranos

One evening Ebs and I went for a walk in Fagagna. In the twilight, we sat on a lawn near the village overlooking the Italian plain. An elderly Jewish couple, a distinguished man with snow-white hair, who had served as professional officer before and during the First World War in the Austrian Army, and his wife, who was much younger with a slightly disturbed look, sat next to us. Following the German occupation of Austria, they had fled from their native city Graz to Trieste and then to Udine. As the situation of the Jews of Italy became worse they moved to Fagagna, where they were taken in by the village priest, who was the only person there who knew they were Jewish.They invited us to visit their room in the priest's house. It was small and simply furnished and it soon became clear that they had no money of their own and lived on the priest's charity. They showed us pictures of their house in Graz, but did not complain about their present circumstances. They wanted to return to Trieste to find out what had happened to their belongings. We gave them some money which they reluctantly accepted and told them which Jewish organizations they should turn to for help. The man gradually became more nervous, and asked to speak to me in private. His wife left the room, and then he whispered to me: "I trust you and feel compelled to confess something that troubles me very much. I was baptized before leaving Trieste as I thought this would help us. Here, in Fagagna, we go to Church regularly. Only the priest knows that we are baptized Jews. As an official and practicing "Christian", I feel bad about accepting your help. Should I return to Judaism?" I felt as if this man was a modern Marrano since like the Spanish Jews during the inquisition who had been forced to convert to Christianity by prevailing circumstances, he still felt Jewish in his heart.

Berches and Shalet

My cousin *Sally* Stern and his wife Erna lived in Paris. After the Germans occupied France, they got hold of false papers and fled to a small village in the Dordogne where no one knew their true identity. They were hired as servants by a wealthy physician - Erna as a cook and Sally as a waiter, chauffeur and odd-job man. Erna and Sally spoke French with an accent. They claimed to have come from Alsace, where a German dialect was spoken. The doctor's son-in-law, also a physician, was a baptized Polish Jew, and one of their close friends, a Jewish engineer, were also posing as "Aryans". One Friday night, they were all invited for a dinner. Erna had prepared a delicious meal, including Berches, a white plaited bread covered with poppy seeds, traditionally eaten by Jews on Friday nights and "Kartoffel-Shalet," a spicy cake made from grated potato. Shalet and Berches are traditional Jewish dishes. After dinner, the son-in-law took my cousin aside and asked in a low voice; "Leon (his assumed name) what do you call the bread you served us"? Isn't it Berches?" Leon: "Never heard that name." Afterwards the engineer remarked furtively: "Leon, the Shalet was really good!" Leon claimed that both were local dishes in Alsace, and denied any knowledge of the Jewish names. How ironic it was that in these terrible times one could not even admit to each other to being Jewish.

Several weeks after the Armistice, our regiment went through Western Germany and then on to Belgium and Holland. Our parade through Germany had a special significance for us - these "cowardly subhumans" as Hitler had called us, were still around traveling through Germany, a country we had fought against and learnt to hate. Our guns and armored cars proudly bore our emblem - a large yellow Star of David, the symbol every Jew had to wear on his clothing in shame under Hitler's regime. The Germans knew our emblem well, and grimaced or even made loud remarks as we passed by. One evening in Innsbruck, Ebs and I were strolling along the street behind two old women. Another member of our unit, a dark skinned Yemenite was walking in front of them. One of the women turned to the other and said: "Holy Jesus, there is a soldier from the Jewish colonial army. We must be careful, they are very dangerous since they want to take revenge on us for what we have done to the Jews." How could we possibly avenge the murder of millions of our people?

Before leaving Austria, we passed a camp temporarily housing Jewish orphans from the concentration camps. The children were very excited when they saw our convoy bearing the emblem of the Jewish Brigade. They went wild. The British officers in our unit understood how we felt and let us stop. The children jumped all over the cars, and we gave them sweets and clothes. We spoke to them in Yiddish, German, Polish, Hungarian and Russian. They told us about some of their experiences. We had come across many terrible things, but seeing and hearing these children moved us most of all - in particular, their hopeless gestures when we asked what happened to their parents. That evening we camped near a small German village. Some of the soldiers were still so furious that they wanted to burn down the hamlet as a reprisal for the children's suffering. It was difficult to restrain them. This was the only time that our unit got very close to taking revenge. But there were cases where some of our people hunted SS men and killed them.

During the entire trip through Germany, I always rode in an armored reconnaissance car. On approaching my hometown, Metz, I told my lieutenant that I would like to separate from the convoy for a short time. With Ebs I drove through Metz, visited my house, my parents former department store, the Cathedral, the Moselle, and also searched without success for old friends. Everything that had seemed so large to me in my youth now seemed to be very small.

Further along the route, near Stuttgart, I made a similar trip with Yehoshua Marks, one of my soldiers, who later became a student of mine and was killed in action during Israel's War of Independence in 1948. In the makeshift townhall, we tried to find out what had happened to Yehoshua's uncle, who had married a non-Jewish woman. The officials were not very helpful when they realized that we were asking about a Jew and former judge. Only after I shouted at them in my rudest German, they became afraid and gave us his name and address. We found his house, and rang the doorbell. It took his uncle quite some time to realize what two British soldiers wanted from him, since he did not recognize Yehoshua, who had been a small child the last time he had seen him.

We went through many towns that had been ravaged. Mannheim, a manufacturing town, was one of the worst hit. On seeing the orderly numbered heaps of bricks outside the bombed factories, I thought to myself, the Germans will soon rebuild this country and become the leading industrial center in Europe. I was right.

Chapter 6
From Palestine to Israel

By August 1945, after being demobilized, I returned to Palestine, which was in turmoil - the British did everything in their power to prevent the concentration camp survivors arriving in overcrowded "illegal" ships from getting into the country. The *Haganah* members did their best to smuggle them in, frequently bringing them into armed conflict with the British. The British then developed even more brutal methods of "hijacking" the "illegal" ships, transporting the survivors to "concentration camps." They searched for hidden arms in our settlements, and we fought back, but tried to avoid "personal" acts of terrorism. Other Jewish underground organizations, however, were convinced that only through terrorism would the British be forced to allow immigration and to give up the mandate. In one well-known incident they blew up part of the King David Hotel in Jerusalem housing the British headquarters. Although warned in advance to leave the hotel, many people were killed in the explosion. The British caught and hung some of the people involved, and in a retaliatory act two British sergeants were taken hostage and killed. In this constant struggle, brutality was the order of the day.

During this period I was able to warn my superiors of a large-scale operation planned by the British against us. In the spring of 1946 I befriended a British soldier who, unlike most of his colleagues, sympathized with our cause. He was working as a clerk at the British Army Intelligence Headquarters. On June 27, 1946, he told me about an extensive operation code-named "Broadside," which was to include a countrywide search for illegally-held arms, a raid on the Jewish Agency headquarters in Jerusalem, which was our unofficial "government." He also gave me a list of several leading Jewish politicians they intended to arrest.

With the intention of surprising the Jews, "Operation Broadside" was scheduled for June 29, the following Saturday (the Jewish day of rest). On that day, which went down in history as "Black Saturday," owing to my alert, all secret documents had been removed from the Agency's building. But some political leaders were arrested. To this day, the underlying political reasons why the politicians allowed themselves to be arrested in spite of my warning remains a mystery to me.

The situation became worse from day to day with constant attacks, often with official or unofficial British help. One day a van full of explosives was placed by Arabs and British deserters in the middle of Ben Yehuda street, a main Jerusalem thoroughfare. Its explosion completely shattered the street. I lived close by and was thrown out of bed as a result of the blast. I immediately rushed to the site of the explosion. The street was in ruins. I found friends of mine who lived on the second floor of one of the badly-damaged houses, sitting on the floor in a state of deep shock. Eventually after shouting at them for some time, I got the family to move out of the house. As I was throwing some of their belongings down from their apartment to the street, the balcony began to fall apart, and I just managed to get out before the building collapsed.

On one occasion several of my colleagues in the *Haganah* were caught carrying arms by a British police patrol. They were intentionally released by the British into the midst of an Arab mob and were murdered. At the same time Joel, the only son of my sister Erna was serving in the *Haganah* under my command. On the 28th of February, 1948 I gave him a one-day leave to visit his parents. Their house was opposite a British look-out post. For no apparent reason the British started shooting into the street, and wounded a passer-by, a Hebrew University lecturer. He shouted for help. The shooting continued and prevented the ambulance from getting through. Joel ran down into the street to help, and he was then also shot by the British, and died the next day. The Arabs dominated the narrow gorge of Bab-el-Wad, later Shaar Hagai in Hebrew, which was the only access road to Jerusalem from the coastal plain. They shot freely at the passing vehicles, destroying many. This made the situation of Jerusalem particularly acute. The road leading to the Hebrew University campus on Mt. Scopus passed through an Arab residential quarter and the bus to the University was often prey to Arab attacks. At the time as one of the privileged few to hold a licensed pistol, a long barrelled Mauser, I was assigned the task of serving as armed escort for the University bus and for buses running between Jerusalem and Tel-Aviv. From 1946 to 1948, this kept me fully occupied, and as a result, I had very little time for teaching and research.

One of the few bright spots during this difficult time was my first trip to the U.S. The University had decided to build a Biology Building on Mount Scopus, and together with the Zoologist George Haas, I was sent to the United States to look at laboratory equipment in the up-to-date laboratories there. We traveled in what had been a

"Liberty" ship, hastily constructed during the war for soldiers and arms, subsequently redesigned to carry passengers after the war. We left Haifa on January 4, 1947 arriving in New York on January, 23 after a very leisurely and relaxing trip.

Both Haas and I also wanted to put our trip to scientific use, and since his area of specialization was snakes, he was carrying with him a few specimens that he had not succeeded in identifying in a large milk jug. At New York City Harbor customs, officials asked him what the jug contained. Haas answered "Dead snakes." The custom official didn't believe him. Haas said: "If you don't believe that there are snakes inside just take the lid off the can and see for yourself." As the official did so, some of the densely packed snakes "shot" out, looking very much alive. He must have thought that we were mad, and it took a long time to convince him of the validity of Haas's scientific endeavors.

On the return trip, Haas' snakes caused a sensation that was reported on the front page of the "New York Times." This time he was carrying several live non-poisonous snakes in a large cardboard box. While taking leave of his relatives he left them on the deck. A passenger unknowingly sat on the box and the snakes were released all over the deck. Terrified, the passenger screamed for help, and the fire brigade and the police were called. They picked up several snakes and threw them overboard. Haas was naturally furious when he realized what was happening to his precious snakes.

In New York I lived with my sister Liesel, who had left Germany in 1938. She was working as a children's nurse with the Mahler family, relatives of the composer Gustav Mahler. She looked after their children and grandchildren, and now, after 48 years, she is still with the family.

While in New York the course of my life was changed. My sister Erna, her husband, and my mother had managed to get out of Germany in time, and were now living in Jerusalem. Erna had asked me to visit Alice Ambach, a childhood friend of hers from Metz, who was living in New York. For three weeks I forgot my promise. On February 13, 1947, a date I will never forget, I finally remembered it and phoned Alice Ambach. Some friends were coming to visit her that day and she invited me to join them. As I opened the door into the salon, I saw a young blond girl. Seeing her I felt bewitched, staring at her. I was overwhelmed by an indescribable feeling of bliss. I knew for certain that this girl was my "anima," the woman whom I was waiting for all my life.

"By chance," Lieselotte Wolff, who later was to become my wife, had been invited by Alice Ambach on the day that I phoned. As a famous philosopher put it: God speaks to man through the so-called "chances".

Three days after I met Lieselotte I took her out to a club, and while we were dancing, I said: "You know, I am going to marry you." She had just been divorced and was not ready to try again so soon, but a year later, we did get married.

Haas and I stayed in New York until March 17. During this time I visited botanists and plant physiologists at Harvard, Yale, Columbia and other universities. I was very impressed by the high standard of the research and by the modern equipment they had at their disposal. It brought home to me how the center of the scientific world had shifted from Europe around the turn of the century to the United States. By the mid-20th century the large numbers of German scientists who had settled there when they were driven out by Hitler no doubt contributed to this.

I was well received by my colleagues and invited to give several lectures. Being used to a "formal" educational framework in Germany, I was very surprised to find Professors and Ph.D. students calling each other by their first names.

I also visited some of the botanical gardens on the East Coast. They seemed superior to those I had visited in Europe - not only by being larger but by better presentation, and serving as independent research centers.

In New York, I took advantage of the rich cultural life which reminded me of Berlin in the twenties, before the rise of Hitler. I spent hours in the Museum of Natural History, where I saw dioramas for the first time. In contrast to European museums, plants and animals were exhibited in showcases containing replicas of their natural environments. New York fascinated me. I felt so free on 5th Avenue, that once while sightseeing with Lieselotte I proposed to her, adding: "If you say no, I am going to kneel down in front of you and carry you over to the other side of the street!" As I did this, the passers-by looked on smiling. Liesel showed me the city in all its many aspects - it seemed to be an incredible conglomeration of people from all over the world, and in downtown Manhatten, I found the man-made landscape of mountains, gorges and valleys very exciting. Liesel said that I looked at everything with painter's eyes. Sure enough, from then on I took up painting as a hobby, although I had never drawn a straight line in all my life.

Haas and I wanted to travel across the United States by car, and since my Palestinian driving licence was not acceptable, I had to take a driving test. Accustomed to the traffic in Jerusalem, where very few people had cars at that time, I was very nervous when I had to drive down the narrow downtown streets during my test. The examiner reassured me, telling me that he would make sure that nothing untoward happened. The way in which he talked to me calmed me, and I passed the test with flying colors. Are the people in New York still so helpful today?

We started our cross-country trip on May 17, 1947. Since Haas had no driving licence and I was afraid to drive alone, we took Haas' sister-in-law Maude with us to help with the driving. We bought a used car, a somewhat risky business for those unfamiliar with the American business world. Lieselotte, being a good business woman, helped us in this venture, and on our return to New York, we even sold the car for a profit.

We decided that it would be cheaper to buy our own food rather than use restaurants. In the grocery store, I had my first lesson in American English. Asking for what we called "tinned" food and "tin opener" in the British army, the salesman looked at me uncomprehendingly. When I explained what it was for, he said "For God's sake, why didn't you say straight away that you wanted a 'can opener'?" This was one of the occasions I noticed how far "American" English is from "British" English. One of Lieselotte's experiences in a New York restaurant illustrates this point well. As she helped an old English lady to get through a revolving door, she noted Liesel's accent and complained to her: "I'm so unhappy here, because these people speak the most horrible English. For tomato they say "tomeyto", for vitamins they say "veytamins." What has happened to the English language?" Liesel told her not to worry since the Americans speak American and not English. Shaking Liesel's hand the old lady said: "Girl you saved my life. Now, I don't have to worry any more. Of course, these people don't speak English."

In order to get to know the country well, we decided to drive to California by one route and return by a different one. We only drove between 250 and 350 miles a day, to give us plenty of time to observe the plant and animal life closely and get a taste of American culture. Wherever possible, we drove on small side roads, avoiding the main highways.

On the second day of our trip, the car broke down somewhere between Gettysburg and Frederick in Maryland. Since we had no idea what to do, we waited at the roadside for a while. A farmer riding a tractor stopped, looked at the engine, and then took us to the nearest garage, where we were told that the car would only be ready the next morning. The farmer took us to his house, his wife prepared a delicious meal for us and invited us to spend the night with them. On finding out that we were Jews from Palestine, the farmer and his family, who as deeply religious Christians believed in the second coming of Jesus in Palestine, said: "You really come from the Holy Land, where our Lord lived and preached? Then you are doubly welcome in our home." We talked for hours on end, felt like members of the family. The farmer refused to take payment, and gave us enough food to last us for a few days when we left. This is only one example of the hospitality and helpfulness we met among America's rural population.

One rainy, misty evening we went through a forest of swamp cypresses (*Taxodium*) near Maysville in South Carolina. These huge trees up to 3,000 years old, have strange knee-like roots protruding out of the ground. Hanging down from the branches, streaming with water there is moss-like vegetation. Although commonly known as "Spanish Moss" (*Tillandsia*), this is an epiphytic flowering plant, which takes in water and nutrients from the air. Seeing this type of plant for the first time was a fascinating botanical experience for me.

As it was spring, the desert and semi-desert areas in New Mexico, Arizona and California were in full bloom. Being familiar with deserts in the Middle East and North Africa, on the one hand, I felt at home here, whereas on the other hand, I was confronted by a great profusion of various bizarre cacti - round, elongated, and snake-like - which I had not seen before. I was filled with awe by the column-like saguaros (*Carnegia*), which towered above us, reaching up to 15 meters in height.

During our stay in the United States I was invited to lecture at eighteen universities. The main topics concerned the Adaptations of Desert Plants to their Environment, Plant Geography of the Middle East, and Germination-Inhibiting Substances. Haas also lectured in his areas of interest. I was paid $100 for each lecture - a handsome sum in those days, which boosted our dwindling travel funds. Although Haas was basically good-natured, he was also a "Graunzer" (a moaner), and complained bitterly because I had more lecturing engagements than he did. Consequently, our relations during this long trip were not always idyllic!

During our one-week stay in Los Angeles, the first people I visited were Hermann and Paula Wolff, Liesel's parents. I introduced myself to her mother, mentioning that her daughter had written to her about me. She stared at me in disbelief: "You have no beard and side-locks!" Apparently Lieselotte had described me as an elderly, Orthodox Jew from Palestine. Liesel later told me that she wanted to play a trick on her parents, who disapproved of most of her friends.

As far as my scientific future was concerned, the lecture I gave on germination-inhibiting substances at the California Institute of Technology (Cal. Tech.), was the most important for me. After the lecture, James Bonner and George Beadle, who later received a Nobel Prize, asked me to write an article for the *Botanical Review* on that subject (26). More important, Cal. Tech., one of the leading research institutions in the United States, extended an open invitation to me to spend a year working there.

During our stay in Los Angeles, I lectured at U.C.L.A. where I made many "scientific" friends. On a visit to the Salinity Laboratory at Riverside, a world center in research on salt tolerance in plants, I learnt a great deal from my long discussions with scientists there. This topic was of particular interest to me since desert soil and groundwater are highly saline. The amount of salt tolerated by cultivated plants is also of great practical importance. At present, several of my colleagues at the Hebrew University and other research institutes are still working, and making important contributions in this area. I specially mention Joel de Malach from Kibbutz Revivim in the Negev who in his research station, in collaboration with people from various universities, has shown that certain crops produce high yields when irrigated with highly saline water.

En route from Los Angeles to San Francisco, we went past the remains of what had once been the expansive coastal Red Wood forests. These densely-packed monumental trees, as straight as rulers, some of which reached over 100 meters in height, are among the tallest trees in the world. In my view, San Francisco is the most beautiful American city we visited. After the network of sprawling suburbs of Los Angeles, the elegant simplicity of San Francisco was very impressive.

I also lectured at Berkeley and Stanford University. Often, after my lectures, I was grilled for hours on end about recent events in Palestine, Zionism, the right of the Jews to the country, relations with the British and the Arabs, and my experiences with the Jewish refugees during the War. I welcomed these discussions especially on realizing how badly informed most of the students were. I particularly remember a lively discussion in the presence of several Arab students on one occasion.

From California, we went on through Nevada, Utah, Wyoming, Nebraska, and Iowa. I then stayed in Detroit, where I had family, for a few days, while Haas and Maude continued touring the country. Three uncles on my mother's side had left Germany and settled in Detroit in 1883 - two of them were "black sheep," good-looking and carefree they had lost a lot of money, and were sent to the United States to get them out of trouble. Out of love for his brothers a third son who was not involved in their misdeeds, went with them to the United States. The "black sheep" had written promissory notes forging signatures of their father Benedict (Bendix) Loewenstein. These almost forced him to sell the family house and lands. As my grandmother Sarah wrote in her diary, which is in my possession: "After 1883 we suffered a great deal on account of two of our sons. The fear, grief and distress we experienced are indescribable. Had our fourth son Selig and his wife not helped us, our home and birthplace would have been lost." My uncle Selig who remained in Germany, was a steady businessman and good son. Ironically, as a result of hard work, the two "black sheep" became multimillionaires, whereas the third "good" one remained a pauper.

At the time of my visit, my uncles were no longer alive, but at a family reunion, I got to know the Loewenstein clan, 18 cousins, not counting their children. One of my cousins apparently inherited the gambling spirit from his father - he took me to his country club, where he lost $1,200, which to me seemed a tremendous sum of money, in a poker game.

We arrived back in New York on May 10. A few days later, I left the country in an old freighter destined for Lebanon. During the three-week long journey I met three Jesuit padres, Professors of Anthropology from American Universities, on their way to Lebanon to carry on research on the recently discovered remains of a prehistoric man. We had long discussions about biology, evolution, religion, Judaism and Christianity, Jesus and God. The unconventional interpretation of the principles of Catholicism, which only became clear to me much later after reading the works of Teilhard de Chardin, their breadth of knowledge, sharp reasoning, acumen and understanding of psychology, never failed to amaze me - I understood that the Jesuit order (Societas Jesu), is the intellectual elite among the Catholic clergy.

By the end of May 1947, I was back in Jerusalem. During my absence the conflict between Jews, Arabs and the British had become more acute. Attacks and murders of Jews were daily events, and Palestine was in a state of anarchy. On November 29, 1947, the United Nations voted to partition Palestine into two states - one an Arab and the other Jewish. The Jews accepted the UN decision. Although the proposed Jewish state was very small, the Arab States on the other hand, were vehemently opposed, and threatened to destroy it by every possible means. Nowadays, the world has conveniently forgotten that it was Arab refusal to accept partition that created the so-called "Palestinian problem." The British declared that they would not take responsibility for implementing the U.N.'s partition plan: their mandate over Palestine would come to an end on May 15, 1948, and all their military and administrative personnel would leave the country on that date.

Despite the uncertain future, the 29th of November 1947 was a day of rejoicing among the Jews in Palestine. After 2000 years, there would be a Jewish State once again. We danced in the streets of Jerusalem, and many British soldiers, who were pleased to be going home, joined in. The British armored cars with soldiers laughing and singing and children singing the Hatikvah, the national anthem, sitting on top, drove through the narrow streets. The bloody enmity between the Jews and the British seemed to nearing its end. As I danced and sang, black thoughts clouded my mind. The struggle between Palestinian Arabs, who rejected the Partition Plan, supported by the neighboring Arab countries and the fledgling Jewish State had already begun.

As a member of the *Haganah*, I knew only too well how badly armed we were. How could we fight the armies of Jordan, Egypt, Syria, and Iraq, as well as the Arab irregulars? How many of those out in the streets would be killed within the next few weeks. My worst fears were realized on that same afternoon when Arabs set fire to a Jewish business quarter near the Old City, and the British police wishing to remain "neutral", did not allow us to come to their aid. During this period, I was exclusively involved in *Haganah* work. Together with my friends Richard Richter and Gassner, a garage owner, I "made" armored cars by putting special plates on civilian trucks. We produced primitive hand grenades. On the *Haganah's* order, I tried to persuade Arabs living outside the Old City in mixed Arab-Jewish quarters not to desert their homes. At that time, I was living in a house belonging to Ibrahim Kattan, a Bethlehem-born Arab, in Talbieh. Over the years we had become close friends. A few days after November 29 he told me that the so-called "Arab Higher Council" had ordered him to leave his house in Talbieh and return to his other home in Bethlehem. The same order was given to the other Arab residents of Talbieh. After the defeat of the Jews by the Arab armies they could return to their homes. Kattan was convinced that the Jews would either be killed or driven out of the country and as a good friend, urged me to leave while there was still time. He even offered to store my valuables for me, until I could come back after the Arab "conquest" of Palestine. I implored him not to leave his house, explaining that the *Haganah* would protect all the Arabs of Talbieh. I could not convince him. The next day he had left his house. After the Six-Day War in 1967, I visited Kattan's house in Bethlehem, but he was no longer alive. Today, the Jews are accused of creating the terrible Palestinian refugee problem, and although there were some cases in which Arabs were driven out of their homes and villages, this was certainly not the policy of the *Haganah.*

The situation of the Hebrew University on Mount Scopus became almost untenable after November 29. Arab attacks on the bus carrying students and teachers to the University and doctors and nurses to the Hadassah Hospital, also on Mount Scopus, occurred daily as it passed through the Arab quarter of Sheikh Jerrakh. Many people were wounded and killed.

Under the circumstances, it was decided that it would be safer for the buses to travel in convoys. On April 13, 1948, the Arabs managed to stop one of these convoys by blocking the road with a stone barricade. The buses were showered by a hailstorm of bullets and hand grenades, and then the Arabs came down to continue the attack, setting fire to the buses. Seventy-seven staff members of the Hebrew University and the Hadassah University Hospital were murdered within sight of a British military post. The British, who observed the event, refused to interfere or to allow the *Haganah* to come to the aid of the convoy, ignoring the pleas of Dr. Yehudah Magnes, who was then president of the Hebrew University.

I was out of Palestine during this period. In February 1948, Dr. Magnes, Dr. Werner Senator, Senior Administrator of the Hebrew University, and Dr. Manka Spiegel of the Department of Organization and Information sent me on a fund-raising mission. The University budget was entirely based on voluntary donations from Jews all over the world, collected through the "Friends of the Hebrew University," which had branches in 36 countries. Their main office was in New York. The World Zionist Organization and the Jewish Agency only contributed small amounts, and these were stopped in 1947, in the wake of the urgent need to channel this money towards housing and the general absorption for the thousands of new immigrants. On behalf of the University, I was sent to countries in North and South America, since over 80% of the University's budget came from there. The *Haganah* agreed to release me since I was already too old for active service, but asked me to carry out certain confidential tasks for them while in North and South America.

On February 29, 1948, I took leave of my mother, sister and brother-in-law. Although I wanted to delay my departure because of my nephew's recent murder, the family insisted that I go as soon as possible because of the urgency of my mission. With a heavy heart, I agreed.

The trip from Jerusalem to Tel-Aviv in one of the homemade armored vehicles was difficult. At Sha'ar Hagai we were shot at as usual. It was a very strange feeling to sit back passively as a "bloody civilian" and not carrying a gun. We arrived safely in Tel-Aviv, but the car behind us in the convoy was damaged, with several people killed and wounded. On March 2, 1948, my plane took off from Lydda. As the lights of Tel-Aviv disappeared, I could not help wondering whether I would ever return to a Jewish State.

This was my first experience of a flight in a civilian airplane, although I had flown sometimes during the war with a bomber pilot who had taken me for a short "spin." Planes flew much lower than today, allowing the changing landscape to be followed. I particularly enjoyed the passage over the mountains of Corsica, and seeing the change in vegetation as we went from the Mediterranean into Central Europe. The plane stopped in Rome and Amsterdam, where I managed to visit a plant biochemist working on the chemistry of germination-inhibiting substances. Although I only knew of him through publications and from an exchange of letters, he received me like an old friend.

The next stop was Glasgow. Soon after landing, the following announcement was made over the loudspeaker: "Passenger Evenari from Palestine is requested to come immediately with all his luggage to the airport police station." I was coldly received, they took my Palestinian passport, scrutinizing it carefully, everything had to be unpacked and checked and rechecked. I then had to undress and go through a very unpleasant body search, including my mouth and anus. They were apparently looking for secret papers. After the search, I was interrogated. I told them that I was a botanist from the Hebrew University and had served for five years in the Royal Artillery as a

Sergeant Major. They then became friendlier and revealed that I was suspected of being a Jewish terrorist. I insisted that I was on a mission for the Hebrew University, which they could verify from my papers. Since I was not carrying any *Haganah* papers, they did not find any incriminating evidence on me. They still would not believe me, and asked me to give them some names of English botanists. After making a few phone calls, they let me go. The whole palaver had taken two hours, but I still got my plane as the police had ordered the pilot to wait for me, which I thought was decent! When I entered the plane I was met by hostile looks from the passengers who were apparently asking themselves for what kind of crime I was retained for so long.

Our next stop was in Shannon in the Republic of Ireland. On showing my Palestinian passport there, I was treated like an old friend. The controller called over all his colleagues, we shook hands and I had to drink with them a whole glass of Irish whisky to the toast: "To the Jews of Palestine, who fight the f--king British" and "Boys, give it to them." At that moment, I thought of our Irish major in the 1. Pal. Light A.A. Bty., who had vowed to come and help us fight the "bloody English."

On March 5, after a 72-hour flight I arrived in New York, where Liesel was waiting for me at the airport. I stayed in the United States until May 15, working day and night. I suppose I wanted to overcome my feeling of guilt because I was physically safe, while back home, everyone was in danger.

With regard to the *Haganah* work, I cooperated with Aharon Katschalsky, a former student of mine, who was then holding a senior position in the *Haganah* in New York. A physical-chemist, he was working in the field of thermodynamics, and was definitely in the "Nobel Prize Class." Later he was murdered in Ben Gurion airport by Japanese terrorists on May 30, 1972. Once in the subway we were discussing *Haganah* matters in Hebrew, naively believing that we were speaking a "secret" language. Suddenly, the man sitting next to me said in fluent Hebrew: "I would advise you to be more careful in the future. There are many people in New York who understand Hebrew." We were very embarrassed and took his advice into account.

The most important part of my mission for the Hebrew University was to visit influential and rich people to explain the University's desperate situation to them and appeal for their immediate help. According to my diary, I visited nineteen people, including Eugene Untermeyer, Robert M. Strauss, Walter E. Meyer, Maurice Wertheimer, and the Warburg Family. Although it was not always easy to get appointments with these busy people, I usually succeeded in achieving my aims. I flew to many cities, often only for a few hours. In Albany, Northampton, Syracuse, Chicago, Milwaukee, St. Louis, and Buffalo I spoke at meetings of local chapters of the "Friends of the Hebrew University," and other large Jewish organizations, such as the "National Council of Jewish Women" and the United Jewish Appeal. I tried to convince them that they should pay special attention to the pressing needs of the Hebrew University. I also spoke at gatherings in private homes, usually of "Friends of the Hebrew University." My speeches were followed by question-and-answer sessions, going on for hours. After these meetings, I would feel completely exhausted, particularly if I had more than one meeting a day. In addition, I was interviewed many times by radio and newspaper reporters, who were interested in eye-witness accounts of what was happening in Palestine.

During this period, I only had one moment of great personal happiness - on the morning of April 6 1948, I was married to Lieselotte by the justice of the peace John Muller in Hoboken in New Jersey. When I took Lieselotte by taxi from her flat I was not completely certain whether she would come with me. We didn't have time to celebrate our honeymoon since I had a press conference and another meeting scheduled for that day. We did not tell anybody about our marriage, but from then on, Liesel helped me with everything, except for the *Haganah* work. We were now a team.

Initially, I had some difficulty in persuading Liesel, who had decided never to travel by air, that unless we did so, I would never manage to fulfill my schedule. I had to go to Rochester to speak, about a ten-hour journey by train, so after much deliberation, Liesel agreed to take a plane with a stop-over in Albany. During take-off, a tire exploded and the plane limped slowly back to the airport. The passenger sitting next to Liesel said: "We're lucky that this did not happen when we landed, we would have made headlines then." We returned to the airport terminal where we were supposed to wait until the tire was replaced. However, Liesel refused to go back into the plane and we continued our journey by train. Later, Liesel got over her fears and we flew many thousands of miles together during the rest of the trip.

Three days after getting married, I was invited to give a lecture at Smith College, a famous and rather exclusive women's college in Northampton, Massachusetts. I was also supposed to give a talk at the local synagogue. Professor Blakeslee of Smith College invited us to stay at his house. His work on fungi was well-known (the fungus

Phycomycus blakesleeanus is named after him), and he had been invited by the college to open an experimental station for fungus research after his retirement from academic life.

At breakfast, one of his students came and asked if she could speak to the Professor privately. When he got back to us, he was visibly embarrassed and asked me to come down to the cellar to see the heating system. I did not know why he was acting so strangely, but felt there was something wrong. He hesitantly told me, not daring to look me in the eyes, that the girl who just had spoken to him was the daughter of a cousin of my ex-wife Alice. Apparently, Alice wanted to come and live in America, and was being held at Ellis Island; she had called her cousin to track me down, in the hope that by declaring myself as her husband, it would help her to obtain an immigrant permit. Her cousin had found out that I was at Smith College, and sent her daughter to ask me to help Alice. The reason for the Professor's discomfort finally became clear - he must have thought that I was either a bigamist or that I had introduced my girlfriend as my wife. Fortunately, I was able to prove that I was divorced from Alice, and that Liesel and I were legally married. I phoned my ex-wife's cousin, Dr. Kurt, to explain the situation. She was furious and vowed not to do anything further to help her cousin. Later, Liesel visited Dr. Kurt to ask her to forget what had happened and to help Alice to get into the country.

On the morning of May 15, 1948 we heard the news over the radio of the declaration of a Jewish State, which Ben Gurion had named Israel. After 2000 years without a country of our own, we felt deeply moved and grateful to be part of this miracle, but regretted that we had not been able to be in Israel for this historical event. Shortly afterwards, the radio announced that the armies of Jordan, Egypt and Syria had invaded the new Jewish State. Although I had expected this move, it dampened our spirits.

Chapter 7
South America

I wanted to return immediately to the new State of Israel, but was persuaded that I would do more by continuing my mission. Consequently, at midnight May 15, Liesel and I boarded a flight to Argentina with stopovers in Trinidad, Port of Spain, Belem and Rio de Janeiro, which took 60 hours. En route from Trinidad to Belem I had my first glimpse of a tropical jungle - huge tightly packed trees forming a seemingly endless ocean of dense gigantic trees.

Extremely tired, we arrived in Buenos Aires and went straight to the hotel. An hour later, the committee of the Argentine Friends of the Hebrew University came to meet us, and to discuss details of the schedule for our five-month stay in South America; and I tentatively agreed, stipulating that I would like enough free time to meet with the eminent botanists and to go and see the main vegetation types there. Although they mostly kept to this promise, I sometimes had to fight for my rights!

The next day, I went to several major academic institutes in Buenos Aires, meeting L.R. Parodi, Arturo Burkart, A.L. Cabrera, A.E. Ragonesse, A. Castellanos, J. Morello, E.M. Sivori, J.P. Grünberg, among others. I culled information about the country's vegetation, found out what books to read, and the main places to visit. Toward the end of my stay in the country, I returned to Buenos Aires with a large herbarium, where my colleagues helped me identify some of the plants, and sent the collection on to Jerusalem.

I was invited to speak at many institutes and universities. I will never forget the first of these lectures on germination-inhibiting substances at the University of La Plata. At that time, my only knowledge of Spanish came from a Spanish textbook I had hastily read during the flight to Argentina. However, when the customs inspector at the airport spoke to me in Spanish, I did not understand one word. I wanted to give my first lecture in Spanish so I asked someone who worked at the "Friends'" office, who knew German, to translate my lecture into Spanish. I intended to read the text, and towards the end practiced reading the Spanish text many times with the translator.

The day before the lecture three professors from La Plata came to see me in order to finalize the transport arrangements. Afterwards, I realized that I did not feel well, and found out that I had a temperature of 41°C (104°F). Liesel called a doctor immediately, and then both advised me to cancel the lecture. However I felt that when the people back home were engaged in a life-and-death struggle, the least I could do was to fulfill duties I had taken on.

The next morning, still retaining a high temperature, we drove to La Plata. The lecture room was crowded, with many students standing in the corridor. That an Israeli was to speak there, probably accounted for the large audience, rather than the subject of my lecture. After reading the first words from the written text in a kind of feverish trance, I put my script aside and began to speak freely in Spanish, losing all my inhibitions. From then on, I felt that I really knew Spanish and lectured often in that language.

After the lecture, the audience responded with thunderous, seemingly unending applause, which sounded to me as if it was coming through a layer of cotton wool. The students were accustomed to hearing foreign lecturers in English, and apparently appreciated a foreign scientist speaking their mother tongue. For two weeks afterwards, I was seriously ill.

In Argentina, I spoke in German as well as in Spanish since a significant proportion of the community consisted of German Jews who had come to live there after Hitler's rise. I also gave some lectures in Yiddish to Jews from Eastern Europe. I did not learn Yiddish at home as a child, but took lessons in Jerusalem. Since it is a mixture of medieval German, Hebrew and some words of Slavic origin, I found it an easy language to pick up. These lectures were also extremely well received even though my Yiddish was full of mistakes. By speaking Yiddish instead of German, one of my mother tongues, I did not run the risk of being booed off the stage, as happened to a Professor of Geology, also from the Hebrew University, when he spoke German to his audience, who hated Germany and everything associated with that country. Afterwards, when asked where I had picked up such a strange Yiddish dialect, I had to admit that I learnt it in Jerusalem!

My public activity on behalf of the university began the next day with a press conference attended by many journalists, who asked about current events in Israel and my personal experiences there. Subsequently, I visited leading Jewish and non-Jewish personalities, giving countless lectures in synagogues and talks to the major Jewish organizations, and many more press and radio interviews. As well as Buenos Aires, we went to other cities with large Jewish communities, including La Plata, Rosario, Cordoba, Santa Fe, Mendoza, Tucuman, Santiago Del Estero and Posadas.

I promised Liesel, who attended all my talks and lectures, that I would always try to vary the themes somewhat so as not to bore her. My severest critic, she claims that I kept my word.

The hectic pace of our trip is well illustrated by our itinerary in St. Louis. We traveled there by overnight train, sleeping little. At nine the following morning, we were met at the station by the local "Friends" Committee. The Chairman took us to our hotel, where we only had a few minutes to ourselves before a press conference, lasting two hours. Then came a luncheon during which I gave a talk, a radio interview and a sight-seeing tour. Once back at the hotel, we barely had enough time to dress for a cocktail party, which the local community held in our honor, followed by a large dinner, attended by many leading Jewish communal figures, where I presented my standard speech about the Hebrew University. By then, it was time to pack and leave for our next destination.

On one occasion, I nearly got myself into political trouble. I was speaking at the largest Jewish sports club in Buenos Aires about Palestine and the new State of Israel, and mentioned that in the Declaration of Independence proclaimed by Ben Gurion, freedom was guaranteed to all citizens regardless of race or religion. Then I added: "Ustedes aqui en Argentina son tambien *mas o menos libros.*" (" Here in Argentina, you are also more or less free".) The audience fell into dead silence, and one could have heard the proverbial pin drop. After the lecture, several club board members drew me aside and explained that I had said something dangerous. At the order of General Juan Domingo Peron, who was Dictator of the country at that time, secret service men on the look-out for revolutionary statements, came to all public gatherings. Apparently "Mas or menos libros," could have easily been interpreted as concealed criticism of the government. Fortunately, my thoughtless naive remark had no negative consequences. It may, however, have been one of the reasons behind an invitation to a short private audience with Peron, at which his wife, Evita, a very beautiful woman, was also present. He asked me about the situation in Israel and our struggle with the British, for whom he apparently had little affection. Evita took part in the conversation, and I had the distinct impression, as did most of the people who worshipped her, that she was the real power behind the throne. I mentioned that, as a botanist, I would like to see the local vegetation and in particular the Missiones jungle, and Peron promised to arrange a trip for me with Juan Schwarz from Santa Pipo, who was of Swiss origin and spoke German, as guide. That excursion was the botanical highlight of our visit to Argentina.

In all I gave forty-six talks to Jewish audiences on the importance of the Hebrew University, the new State and its fight for survival, a series of 14 lectures at various universities and also called on many important people to appeal for donations to the Hebrew University. Despite my parallel activities on behalf of the *Haganah,* I had a bad conscience about not being in Israel, especially when the news of the siege of Jerusalem and the occupation of the Old City of Jerusalem by the Arab Legion reached me. Many times I wanted to return but my superiors always put me off by telling me that at the age of 43 I was too old for active army service and should therefore continue my mission in Argentina.

The first area of botanical interest I visited was the Pampas, a steppe with a thick grass cover, extending over an area of 310,000 square miles (two and a half times the size of Israel). Since it is the most fertile part of the country, some of it is used for cultivating wheat and other areas as grazing land for large cattle herds. In Buenos Aires someone told us the old wive's tale that the pampas was so fertile that if one planted a broomstick it would sprout roots.

Back in Buenos Aires, on a visit to one of the largest companies for export of agricultural produce, I noticed that the manager used a large jade-colored stone as a paper weight. Being something of a "rock-hound," I asked what it was. He explained that it was a "wheat-stone", consisting of compressed ashes of wheat burned by the company in order to prevent a fall in the price of wheat. In later years, 1,000,000 tons of wheat producing about 20,000 tons of ash, were burned for the same reasons. This ash formed a compact mass of stone at high temperatures, and was used as paving material. When somebody remembered that plant ash is rich in inorganic nutrients, the wheat-pavement was taken up, pulverized and marketed as fertilizer for wheat fields! The manager kept some of the stones as a reminder of human stupidity! He gave me a piece of wheat-stone, which I kept in my collection, until another rock-hound stole it, thinking that it was a valuable special piece of jade.

Pampas in its natural state presents a botanical conundrum similar to the prairies of North America. Was it always treeless grassland or had it been turned from a forest into a steppe by man? Trees planted around the ranches there certainly flourished - a point in favour of the "forest theory." The first Europeans to visit the pampas described it as a treeless steppe. Perhaps it has always been in this state since it is completely lacking in indigenous trees. Experiments that showed that tree seedlings died if they did not receive special care, since their roots cannot compete with the grasses, which seem to be especially well adapted to the prevailing soil type and climate, further support the "steppe theory."

We came across another example of man's criminal carelessness toward nature during a train journey from Buenos Aires to Tucuman, a University town at the foot of the Andes. We passed through vast stretches of pampas with thousands of grazing cattle and horses and seemingly endless wheat fields. As evening approached, the conductor sealed all the windows with wet cloths and asked the passengers to keep them closed throughout the night, in order to prevent dust from coming into the compartment. Indeed, as we approached the city of Santiago Del Estero the next morning, red dust covered everything in spite of these precautions.

We then left the pampas and went through a semi-desert with vegetation similar to the deserts of North America - opuntias and other types of cacti, creosote bushes and very few trees. The shrubs were well spaced, and the loess soil easily picked up by the wind. It seemed illogical that an area with an average annual precipitation of about 500 mm should be a semi-desert. Another aspect I found puzzling as the large stacks of wood at every station. Where did all of it come from?

Subsequently, during a tour of the area, botanists from Tucuman University explained that the entire region had once been a forest with two commercially valuable species, the *Quebracho blanco* and the *Quebracho colorado,* deep inside it. Beautifully-veined "colorado" wood is unusually heavy and very durable, while "blanco" makes high-quality charcoal. Hundreds of trees had been cut down, laying bare the soil, which then became eroded in the tropical rains, in order to reach the few Quebrachos that grew in this mixed forest. Under the prevailing conditions, the forest could not regenerate, and a semi-desert was formed. Thus, man's folly and disregard of nature had destroyed a valuable forest within a short period.

In Tucuman professional colleagues and the members of the small Jewish community made us feel very much at home. Tucuman, founded in 1565, named after Miquel Lillo, was one of the first towns established in South America. The university there, has an admirable botanical institute, which was headed by Prof. Horatio Descoles at that time. I was invited to give a series of lectures there. During my first talk I made a blunder - for some reason, I mentioned Guayule, an indigenous plant in the deserts of North and South America, which like the rubber tree, *Hevea brasiliensis,* contains latex. The audience froze and thinking that I had pronounced the name incorrectly, I wrote it on the blackboard. Turning round, I said to Prof. Descoles, who was sitting in the front row, "Usted conoce esta planta?" (Do you know this plant?). He jumped out of his seat as if bitten by a tarantula and with a furious expression on his face said: "Si." Later, when Descoles left the hall, the audience burst out laughing, and some of the students "enlightened" me. During the Second World War a fervent supporter of Peron, Descoles had tried to cultivate Guayule on a large scale, claiming that the rubber produced from this plant would bring millions of dollars into the country. However, his scheme failed. That very morning, one of the local newspapers had published a very malicious article relating to this subject on its front page, which was divided into two columns, comparing the Professor's claims and the actual results. For example one part read ironically: "Professor Guayule, in 1944 you said: "Guayule will bring millions of dollars into the country," and the other: "And what do you say now, Professor Guayule?" I had inadvertently mentioned the Guayule plant on the same day as the newspaper attack.

Descoles seemed to understand that my reference to Guayule was an unfortunate coincidence, since he never brought up the subject again, and did all he could to make our stay in Tucuman as pleasant as possible, arranging an excursion, financed by his institute, into the Andes mountains. A jeep was supposed to come to pick us up at our hotel at 9 a.m. Liesel made me angry by coming down 15 minutes late, and then we waited in front of the hotel for three hours. Thinking that there had been some sort of misunderstanding, we hired a horse and buggy, the main means of transport in Tucuman at the time, to take us to the institute. There, they explained that there was no petrol available and so we could not start our journey that day. "Mañana" everything would be all right. This "mañana" lasted for three more days! We had come to understand that the real meaning of mañana is "sometime in the future," and what we called "mañanism" is one of the curses of the daily life in South America. During the three-day waiting period we did some sightseeing. Descoles showed me round the institute's herbarium, which is the largest and most comprehensive in the region. Meanwhile Liesel paid a visit to the small zoo, also at the institute. About two hours later I went there to pick her up, only to find her scratching a puma behind the ears through the holes of a cage like a housecat and licking Liesel with his large red tongue and putting his head into her hand. I did not dare to move fearing that by doing so I would excite the puma. On spotting me Liesel slowly withdrew her arm, and told me how she had first talked to the puma like she did to the many cats we kept at home. A small gray cat occupied the next cage, bearing the label, "wild cat." Thinking the people mad to keep what seemed to Liesel like a small domestic cat in a cage, she talked to it and put one finger into the cage. However, in this case, the cat held her finger firmly between its teeth and claws, only releasing it when Liesel "spoke" to it in her sing-song voice. The keeper confirmed that this was really a wild aggressive cat and told us that this was dangerous play indeed, and could have cost Liesel her finger. He had tried often to tame wild cats but never succeeded.

The Selva, a strip of subtropical to tropical jungle at the foot of the mountains, was our first stop in the Andes. This vegetation type owes its existence to the direction of the windflow over the Andes mountain range: On the Eastern side, so-called fall winds drop from a great height discharging their humidity to give an average annual rainfall of about 1700 mm, which combined with the heat favors the development of the Selva.

People who have no direct experience of jungle vegetation cannot appreciate the real potential of plants to grow and proliferate. The trees were so tightly packed that the sunlight could not reach the soil. Every tree there is a "garden" with an entire ecosystem growing on its stems, branches and even leaves - ferns, orchids, *Bromelias, Tillandsias, Philodendrons,* algae and lichens, which are epiphytes, getting their nutrients from air and rainwater. I climbed up one tree to collect epiphytes, and not including the brightly colored algae and lichens, came back with two orchids, a *Philodendron* with giant leaves, three species of cacti and four types of *Tillandsia.* When I plucked a large *Vriesea* from one branch, a stream of rain water, which had accumulated between the plant's leave poured down onto Liesel, who was standing below.

Another feature of the Selva was new to me. The forests I had come across in Europe, Kurdistan, and North America only contain very few species of trees - usually four or five. In the Selva of Tucuman, there are on average 35 different tree species per hectare.

From the Selva, we went east, driving on a dangerous, winding road which was so narrow that there was no room for over-taking. As we climbed to a height of 4200 m, I counted 28 hair-pin bends along a two-kilometer stretch. The beauty was so overwhelming that I could not help expressing my admiration: "Que paisaje marravillosa!" (What beautiful scenery!). Every time I said this, the driver turned round and nodded in agreement, even when we were passing precipice after precipice. Such behaviour on these steep roads frightened Liesel who told me in German that it would be better if I admired the landscape silently, as we might otherwise end up somewhere down below!

At an altitude of about 2000 m the steep slopes were covered by all kinds of cacti - some tree-like, 6-8 m. high, others in the form of shrubs and some even lying on the soil. Most impressive was the Cardonal, a cactus with a broad stem about 50 cm in diameter, which has sausage-shaped broad side branches protruding at a height of about 2 m. Many of the cacti were decked with *Tillandsias*, covering even their spines. On another occasion, at an altitude of about 3,500 m, where it becomes very cold, I saw ice-covered cacti. This brought home the fact to me that cacti do not only grow in hot deserts. Liesel and I were so fascinated that we jumped off the jeep, and started running up the slope in order to get to one of these plants, but we had forgotten that we were so high up, and after a few steps, gasping for air, we were forced to stop. Further up, there were no higher plants, only multicolored lichens covered the bare stone and sparse soil. We got out of the car at about 4200 m, where we had an awe-inspiring view of the snow- and ice-covered peaks towering above us.

We returned to Tucuman, where we took leave of the Institute. As a personal farewell present, Descoles gave us the five-volume work that he had edited, entitled "Flora of Argentina." Each book, weighing about 10 kg and 50x30x9 cm in size, was magnificently printed with beautiful, unusually large plates, big lettering and ample margins, as was once the custom in similar floras 200 years earlier. The dedication of these books to the dictators ruling the country including General Juan Domingo Peron (volumes four and five) apparently allowed this luxury. This gift put us in an awkward situation as there was no way we could take the 50 kg of books with us by train. One of the local members of the "Friends" kindly offered to send these tomes onto the main office in Buenos Aires, and only many years later were they sent to me in Jerusalem, where they are part of my library to this day.

We then went back to Buenos Aires. From there we went on an excursion to the jungle of Missiones - the climax of our visit. We first flew to Posadas to meet Gerardo Juan Schwarz, our guide. The next day, we traveled in a small motor boat along the Parana River, which forms the border between Paraguay and Argentina to Puerto Iguazu - the meeting point of Paraguay, Brazil and Argentina. On both banks of the river, the jungle comes right up to the water's edge. Lying back leisurely in deck chairs on the slowly moving boat, the many shades of green, the splashes of blue-lilac, pink and flame-red of the tall flowering *Jacarandas*, *Tabebuyas*, and *Erythrinas*, like an impressionistic painting, were a feast for the eyes. Three levels of vegetation were clearly distinguishable: the tallest included trees such as the *Jacaranda*, trees such as *Chrysophyllum* are in the middle range and various leguminous shrubs, mainly mimosas formed the lowest level. The trees were covered with epiphytes and long lianas, hanging from the branches. Between the trees, along the river banks, were dense thickets of giant bamboos, which Schwarz informed us only flowered in synchrony every 30 years.

We arrived in Puerto Iguazu where Santo Bertoletti, a local forester, joined us. Both he and Schwarz knew the flora of Missiones well, and with their help I was able to identify most of the many plants I collected.

Although we already had a foretaste of a jungle in The Selva of Tucuman, in Missiones we were confronted with the tropical jungle, characterized by over 2000 mm annual rainfall. The vegetation was much more luscious, tightly packed and varied. At one point, our guides identified over one hundred different species of trees, ranging between thirty and forty meters in height, within an area of two hectares. They were familiar with scientific names unknown to me. I also took note of their popular Indian names, as I liked the sound of them - "Guatumbu," "Ysapuy," "Baporoiti," and "Aguay." There were also more epiphytes and lianas there. I collected three different species of orchids, ten varieties of fern, five types of fig (*Ficus*) and five plants of the Bromelia family (*Tillandsia, Vriesea, Aechmea, Bilbergia, Bromelia*) from one tree. The vegetation fascinated us - even as darkness was beginning to fall, we were still collecting plants.

At dawn the next morning, we were already on our way to the Iguazu falls which were still in their original state, untouched by man. As I noted in my diary: "One immense roaring wall of water and numerous smaller waterfalls, interspersed with islands of green vegetation. A yellow thundering flood of water shoots down the rocks, and huge clouds of water vapor rise from below in a rapidly flowing stream of water. I found a plant of the *Podostemonaceae* family so firmly glued to the rocks that I could not pry it loose. Its green photosynthesizing roots excrete a substance attaching it to the rocks, and its huge metamorphous shoots, are ideally adapted to this strange habitat. Flocks of birds fly toward the waterfalls, into which they seem to disappear into their well-protected nests in the rocks."

At an altitude of about 1000 meters, there is a sudden change in vegetation - the leafy trees give way to a magnificent forest of coniferous evergreen trees, primarily consisting of *Araucaria*. Bertoletti had a hut in this forest, where he received us warmly and treated us to venison for breakfast. He was in charge of the upkeep of the *Araucaria* forest, which had been severely damaged as a result of irresponsible cutting down of trees. He established a tree nursery, and succeeded in halting the destruction in this unique forest. It was the only successful afforestation project that I came across in Argentina.

Among the *Araucaria* I was surprised to observe tree ferns of the *Alsophila* species, relics of a much earlier period, during which these plants constituted the main vegetation on earth. I also saw Mate trees, which are cultivated on a large scale in Missiones, growing wild. Mate is the national beverage in Argentina and Paraguay but it was not the proper thing to offer this "common people's" drink to foreign visitors in either private homes or hotels or restaurants. We drank it for the first time towards the end of our excursion to the Andes. The day before our return to Tucuman, our driver went through his parent's village, and invited us to their home. His entire family was sitting in a circle drinking Mate out of a calabash, made out of the dried hollow fruit of the Lagenaria gourd, in a large, sparsely furnished room. A bombilla, a small silver tube with a sieve-like end served as "straw". The calabash and bombilla were passed around. The grandfather and the toothless grandmother drank first. Then, as guest of honor, it was Liesel's turn. I did not dare to look at her as I knew that she is so particular about hygiene. Entering a hotel room, she immediately cleans the glasses. However, knowing that by refusing she would offend our hosts, she managed to drink the Mate without showing outward signs of aversion. Then, came my turn and the Mate went round and round many times. It certainly refreshed us, and our fatigue was gone. We understood now why it is such a popular drink in South America.

Once we went past the remains of an old church and several houses in Missiones. Bertoletti and Schwarz explained that this had once been a Jesuit mission center. The name of the province derives from the many missions that once existed in this jungle. This piece of historical information fascinated me so that on our return to Buenos Aires I made a point of learning more. In the late 16th century, the Jesuits in Paraguay and Missiones converted about 100,000 Indians to Christianity, organizing about 50 encomiendas (communal agricultural settlements) for them. Under the guidance of the Jesuit paters, the Indians followed a socialist religious life style, which reminded me of the Kibbutz. This social structure only persisted for a short period, since the Spanish Vice-Roy, who wished to exploit the Indians as slaves, expelled the Jesuits and destroyed all the encomiendas and churches.

On the return flight to Buenos Aires, we experienced a tropical thunderstorm. Immediately after take-off from Posadas the light, antiquated single-motor plane flew into a dark sulfur-colored cloud. A violent storm with incessant flashes of lightening and thunder broke out, and the plane was bounced up and down, and even seemed to capsize. Apart from Liesel and I, everyone was sick. We were scared to death, thinking that our end had come, however, a little later, we landed safely in Buenos Aires.

After leaving Argentina, we flew on to Montevideo, Sao Paulo and Rio de Janeiro. There, we continued our work on behalf of the University. In Sao Paulo we visited Felix Rawitscher, a Jew of German origin who had been a Professor of Botany in Freiburg, Germany. On Hitler's rise to power, dismissed from his post, he went to Brazil, where he was appointed Head of the Botany Department at the University of Sao Paulo, which became a first-class Institute under his direction. His scientific studies at that time focussed on the elucidation of soil fatigue that occurs in tropical soils. Tropical forests only grow in areas of very high annual rainfall. If the soils in these regions are laid bare by cutting down, nutrient salts are no longer taken up by the trees, but are washed away by the heavy rains. Rawitsher explained that within 10 years of such leeching, the tropical soils became completely unproductive. Since then, this particular problem has become more acute and widespread. Rawitscher also wrote the first botany textbook in Portuguese.

In Sao Paulo, we also went to see the "Orchidario" of Professor F.C. Hoehne, a German who had been living in Brazil for many years. The largest and most beautiful collection of orchids I had ever seen, they were all grown in the open under natural conditions. Next to it there was an area of untouched jungle where orchids grew wild. Kolibris and giant multicolored butterflies that we had not seen either in the Selva of Tucuman or in Missiones, abounded. In 1909, Hoehne took part in the first long scientific expedition into the jungle of the Matto Grosso, where he collected over 4,000 new species. He related how many of his companions on that expedition died of scurvy because they would not eat the leaves of wild plants. Hoehne also tried smoking various herbs that the Indians used, which gave him hallucinations in brilliant colors. His book on the medicinal and poisonous plants of Brazil is a classic. It describes among other topics the plants used by the Indians as arrow poisons.

By October 11, 1948, we were back in New York, where Dr. Senator, the Chief Administrator of the Hebrew University, who was there on a visit, received us and brought us up-to-date on the situation in Israel.

Chapter 8
California and Return to Israel

During my absence from Israel, the *Haganah,* and later the combined *Haganah* and Jewish underground army units that formed the Israel Defense Forces ("Zahal"), managed to defend the University on Mount Scopus against attacks by the Arab Legion. Although the road to Mount Scopus was in Arab hands, the Armistice agreement between Israel and the neighboring countries assured us free access to this site, and therefore the possibility of continuing our work at the University. The Jordanians, however, never kept their promise; the Arab legion only allowed us to bring police guards via the access road to the University in closed convoys at prearranged times. In this way, we were effectively cut off from the University premises, housing all our books, instruments and collections.

In the wake of these circumstances a skeleton University was established in 55 buildings scattered all over Jerusalem. The Plant Physiology Section was housed in a former British police building and the rest of the Department in a hut three kilometers away. Most of the students and the younger staff members were serving in the army, making normal research and teaching activities impossible.

In view of these circumstances, Dr. Senator, the University's Chief Administrator advised me not to return to Jerusalem, but to spend a short period working in a research institute in the United States. I immediately informed Cal. Tech. that I would accept their invitation to spend a sabbatical there. They agreed, offering me a Visiting Professorship. We bought a used car, and left New York on January 16, 1949, driving slowly from coast to coast over 20 days. This time, we went by a different route from the one I had taken together with Haas in 1947, in order to see and experience as much of that immense country as possible.

The seven months I spent at Cal. Tech. in Pasadena were among the happiest and most productive of my scientific career. As Anton Lang aptly put it, "No one who has passed through Cal. Tech. has left quite the same person and probably retains a trace of regret at having left" (27). James Bonner had the greatest impact on my research at Cal. Tech. He advised me to investigate the factors influencing the germination of lettuce seeds, focusing on the effect of light of different wavelengths and germination inhibitors. For the next 20 years, this remained the main theme of my research. I was one of the first people to use the phytotron, a remarkable invention of Frits Went that could simulate most environmental conditions ranging from those prevailing in tropical jungles to those in deserts. I became good friends with him. Went was internationally recognized as the inventor of the phytotron, and his name has also gone down in the annals of Botany for being the first to detect the plant hormon termed "Auxin" now known as IAA or growth hormone. His invention paved the way for the study of a new area of Plant Physiology, which also has many important applications in Agriculture.

Through personal contacts and from their seminars and lectures, I learnt a great deal from Went and Bonner. Bonner introduced me to Plant Biochemistry, a branch of Botany I knew very little about previously. I also gained much from my colleagues George Laties, Arthur Galston, Günter Stent and Jean Nitsch, a guest lecturer from France, who subsequently died in a diving accident at a very young age. All of them later held key positions in leading universities.

My scientific education at Cal. Tech. was not limited to Botany, I attended lectures in Organic Chemistry, given by Linus Pauling, who later was awarded two Nobel Prizes, Jessie Greenstein's courses in Astronomy, lectures in Genetics given by George Beadle, another recipient of the Nobel Prize, Minkowski's courses in Astrophysics and lectures in Geophysics and Seismology, presented by Guttenberg, who opened new horizons to me as well as taking a personal interest in my progress. I had the feeling, later on justified, that I was moving in the society of future Nobel Prize winners.

My visit to the Observatory on Mount Wilson with Greenstein and Minkowski stands out in my mind. Using the then new 100 inch reflection telescope, they showed us the moon, Saturn, double-stars, star-clusters, many spiral and ring nebulae and star spectra of various kinds, explaining their scientific importance. In the giant dome, observers stood on a small platform, which raised them to the height of the eyepiece of the telescope. Looking through it, we felt as if we could soar into infinity, and fly free of mother earth into cosmic space.

Knowing that my time at Cal. Tech. was limited, and not having any teaching obligations, I worked on my experiments from early in the morning until late at night. I was no exception in this respect - at all hours of the night, there was someone to be found on the premises. Once, during the Easter holidays on our return from a trip to Los

Angeles, we went into the laboratories at 2 a.m., and sure enough found other people working in the building.

In spite of my intensive working schedule, I still took "time-off" to speak at various meetings of the "Friends," and founded a new Chapter in Pasadena. I also joined my colleagues on excursions into the mountains of the Sierra Nevada, and the Mojave and Chihuahua deserts.

In Spring 1949, we went with Frits Went and his students on a trip to the Mojave desert. Much rain had fallen, transforming the usually barren landscape into a magnificent carpet of pink, violet and yellow flowers. Went, creeping on his belly, searched for seedlings of tiny annual plants, which he jokingly called "belly plants." Liesel, coming across a large, strange-looking lizard with yellow, orange and black tubercules, in her customary manner, approached the animal and talked to it. The animal was fascinated and stood still. Liesel came so near that she could have kissed it. Before she touched it, suddenly, one of the young students took her by the arm and pulled her away, telling her that the creature was a poisonous Gila monster.

Went also organized an excursion to Death Valley, which was also in bloom that year. On the second day, Liesel felt so weak that we both decided to return to Pasadena. On our way back, we tried to work out what was wrong with her. Apparently, she had drunk very little, and I explained to her that in hot deserts one must drink large quantities of water to replace water lost in sweat, even if one does not feel thirsty. From then on, Liesel, a newcomer to the desert, always kept to this rule.

In Pasadena we became the unsuspecting owners of a dog. On her way to fetch me from Cal. Tech. Liesel found herself behind a car that braked suddenly in order to avoid killing a dog on the road. Eventually reaching Cal. Tech. she then found it difficult to park as the dog was running around the car. Exasperated she opened the door and impatiently told the dog to make up its own mind. The animal promptly jumped into the car, sitting down next to her - he had apparently adopted us!

We took him home and phoned the SPCA, only to be told that someone had informed them that their dog, a Welsh terrier, had jumped out of their car as they were passing through Pasadena. We were obliged to place an advertisement in the newspaper for the next few days stating that we had found a dog and giving its description; if there was no response, we could keep the dog. We did not really want a dog at that stage since we knew that we would be returning to Israel within a few months, and it would be difficult to take it with us. However, we soon grew to love the dog, which we called Peggy and decided, in spite of everything to take her to Israel with us.

We left Pasadena on September 30, 1949. Before my departure, Beadle, Dean of the Faculty of Science at that time, asked me to stay on, telling me that there would always be a place for me at Cal. Tech. This invitation was really tempting, especially since I knew that in the fledgling State of Israel the research possibilities would be limited for a long time to come. In Israel I would have to forget about working with sophisticated, expensive instruments I had used in Cal. Tech. With my preliminary results concerning the effects of red, infra-red and blue light on lettuce seed germination I was on the way to a new discovery, which is now accepted as common knowledge. I knew that I could not continue this line of research in Israel. After due consideration, I decided that it was my duty to return to my country - a difficult decision, but one that I have never regretted.

In Pasadena we often entertained visiting professors, Ph. D. students and young scientists from all over the world in our home exchanging opinions, discussing politics and telling each other about our own countries. We once invited our neighbor, a young, attractive, recently-divorced woman who came from one of the Southern States to one of these meetings. By chance she sat next to James Henderson, a black colleague from Cal. Tech. and they conversed animated with each other during the course of the evening. The next morning she came over to thank us for an interesting and stimulating evening. She asked us who the interesting Indian was to whom she had talked. Liesel explained that Henderson, a promising scientist, was a negro from one of the Southern States. She asked us if we were certain. When we said "Yes," she stormed out and never spoke to us again.

We left Pasadena for New York with heavy hearts. During the trip, Peggy sat in the back seat of our open Buick convertible, looking at the landscape as if she wanted to remember it. At each stop she carefully buried the remains of her meals in the ground as if she was building up a food store for the return journey.

In the Sequoia National Park, Peggy had to be kept on her leash. She constantly tried to break loose, to hunt the deer, squirrels and other animals running freely around. She even tried to attack a bear which visited us one night. Liesel had an unfortunate encounter with one of the supposedly gentle does. She gave one doe some grapes, and then another one came to beg for food. The first doe, jealous of the competition turned round and kicked Liesel so violently in the stomach, that she could not breathe for some time.

We arrived in New York on October 20 and embarked on our journey on the Greek freighter "Neptunia" twelve days later. For the first three days we had problems with Peggy. She was so well housebroken that she did not dare to mess up the deck, and finally when she could hold back no longer she used the doormat in front of the purser's office, who was not impressed. According to the shop's regulations, she should have been put in a cage, but the sailors, not wanting to take responsibility of looking after her, overlooked this rule.

In Haifa, my assistant Ephraim Konis came to meet us. He smuggled Peggy out, thereby circumventing a long quarantine period.

Unfortunately, Konis died from a brain tumor soon afterwards. I saw him taking his newly-born baby for a walk in a pram. Noticing that he did not move in a straight line, swaying from side to side, I urged him to see a doctor immediately. I was very worried and a few days later, after an unsuccessful operation, he died.

The transition from the United States to Israel was not merely a matter of a geographical change of scene. Whereas in the United States money could buy anything, in the new State of Israel a rationing system, known as "tzena" had just been introduced, including essential items such as meat, vegetables, and oil. Only bread and codfish fillets, which Israel got from the Scandinavian countries in return for citrus fruits, were freely available. The measures were taken to cope with the huge influx of immigrants following the Declaration of Independence. Thus on May 15, 1948, the new State of Israel opened its doors to survivors of the Holocaust. In addition during this period hundreds of thousands of Jews fled from Iraq, Syria, Egypt, Yemen, Morocco and Libya, and made their way to Israel. Every possible effort went into integrating this seemingly unending stream of refugees into the country, providing them with food, shelter and work. Since our means were most limited, "tzena" was an absolute necessity.

Even so, the newcomers who were initially accommodated in tents and huts, had a hard time. The fact that the number of new immigrants arriving exceeded the total Jewish population in the country during that period, provides a good indication of the scope of the problem. Imagine what would happen if the 50 million inhabitants of France would have to absorb 50 million new people into their country during the course of a few years! Thus, Israel managed to take in 680,000 Jewish refugees between May 1948 and December 1951, treating them like brothers; whereas the world all too often forgets how the Arabs, who fled Palestine during the same period, are kept by their brethren for purely political reasons in primitive camps.

For me, the most difficult *tzena* restriction was to make do with 100-gram weekly meat ration. Once, while in a restaurant, an American tourist, who was exempted from *tzena* as a visitor left a half-eaten steak on his plate. I must confess that I found it very difficult to refrain from asking for his leftovers! In spite of the lack of meat, we got enough animal protein from fish fillets that were usually readily available. Interestingly, physicians claim that people's health was never better than during the *tzena* period! Even so, one soon tires of cod fillets at every meal.

After five years, when the rationing ended, I could not face fish without a general feeling of disgust. In 1954, Liesel and I were in England on a fund-raising trip on behalf of the university. Generally, we were served salmon when invited for dinner in private homes, since our hosts did not know if we kept the Jewish dietary laws and only ate kosher meat. Once, at a dinner party in Scotland, our hostess showed Liesel round the kitchen. On seeing a large salmon there Liesel instinctively said "Oh no, not salmon again!" Taken aback, our hostess patiently explained that salmon was the best and most expensive fish available. Liesel told her the reason for her remark, and as a result, we were treated to excellent steak.

Tzena limitations were also reflected in working conditions. We had only the most primitive means available for teaching and research. Breaking a Petri-dish, pipette, or measuring cylinders was a disaster, because we could usually not replace even these small items. The funds for research were very small. There were no proper lecture halls, so that we taught students in private homes. Despite all these difficulties, we continued research and teaching to the best of our abilities. Under these conditions, there was no way I could continue the work I had started at Cal. Tech., which I deeply regretted.

However, these problems were no more than minor irritations, and we lived in a continuous state of bliss. After 2000 years of persecution and torture, and going through the holocaust, at last the Jews had a State. We were now our own masters, and, after so many bitter fights were able to offer all the Jews a homeland, and would build a model state, which we hoped would be the embodiment of our dreams of social justice, freedom and peace with the Arabs. At that time, we had no doubts at all that we would be able to fulfill all these hopes. Mistakenly, we also thought that the existence of a Jewish State would bring anti-semitism to an end.

My mother, sister Erna, and brother-in-law Gerson Stern lived in Jerusalem at this time. After the tragic death of my only nephew, they experienced through the siege of the city. They stood in line for hours on end in the streets under Jordanian bombardment to receive their daily water rations from urns carried on lorries since the main water line from the coastal plain to Jerusalem had been cut by the Arabs. At that time drinking water came from cisterns on which all Arab houses were built. The provisional town council, headed by Dov Joseph, cleaned and sealed the cisterns in order to keep the drinking water consumed down to an absolute minimum. My family had to leave all of their money in Germany, and my sister, working as a physiotherapist, a profession she had studied in Germany, was the only breadwinner in the family. My brother-in-law wrote a book, "The Balance of the World," which was published in Hebrew but did not bring in much money. At the age of 80, my mother started to keep house - a role that she had never carried out during all the years she spent in Germany.

Our salaries were, and still are, very low, and I earned a fraction of what I would have had if I had accepted the position at Cal. Tech. In addition, owing to the University's financial difficulties, salaries were never paid on time - often several months late. Consequently we bought all our food on credit from shops, which were accustomed to carrying out "business" in this way. Trained to believe that it was a crime to be in debt, Liesel once complained to my mother about the salaries not being paid on time. My mother took her by the hand and said: "My child, in Germany, I was a rich woman, I had many maids and servants and everything I wanted. Today, we are poor; we have very little money, but I am happier than I was there. So don't worry about such a trifle as living on credit."

At the time, we lived in two small rented rooms, sharing a kitchen with the flat-owners. We only had one cupboard - the wooden box in which we had sent over our things from the United States. We also had a table and a few chairs. After a while Liesel started looking for a "real" flat, but without success, since there was a serious housing shortage owing to the large-scale immigration.

There was a large house on a hill near us; the basement there was a large, gaping hole, where members of a Jewish underground organization used to manufacture bombs. One of these had exploded in the flat, killing all the people there at the time and completely destroying the flat. However, the solidly built upper stories had hardly been damaged. Liesel often stood in front of the hole, imagining how it could be turned into a flat. One day, she spoke to the owner of the flat, an Armenian, who witnessed the murder of his entire family by the Turks; this incident is described in Werfel's book: "The 40 Days of the Musa Dagh." As one of the few survivors of the massacre, he had been raised in the German Schneller Orphanage in Jerusalem. He offered to rebuild the flat, if we covered the expenses. In return, we could live in the flat rent-free for two years. Liesel immediately went to the University Administration and made a deal with them. They agreed to pay the restoration costs, and we would pay them rent. If we decided to leave the flat, they would have the right to offer it to another professor.

Following a long discussion with Mr. Matossian, the owner of the house, a contract was drawn up. Liesel then built the furniture for the new flat from materials we had brought over from the United States. The floor of the flat was made of tiles, as is customary in the Middle East. Some of these had been destroyed by the explosion. One of the workmen suggested removing all the tiles with an electric drill, but one of the others who must have been our guardian angel, disagreed, insisting to remove the tiles by hand since some of the undamaged ones were quite attractive. We decided to take his advice, and while doing this, came across a large hole full of explosives. Had we used a drill, we would all have been blown to smithereens.

After settling into our new home, we found out that Mr. Matossian, who hated Jews, had made his offer to Liesel thinking that she was a blond, non-Jewish German. By the time he found out the truth, the contract had been signed. When he invited us to his flat he, together with a friend of his tried to convert us to the Lutheran faith, explaining that only this could save us from damnation. We pointed out that the Jews were not good candidates for conversion, as shown over the last 2000 years. Jews had even resisted conversion during the Crusades and the Spanish Inquisition. They also thought that their Catholic brethren would go to Hell! "Let us suppose then, that we will go to Hell because we are obstinate, but what about natives, living in the jungle, who have never heard of Jesus Christ?", Liesel asked. They replied that their fate would be similar since they should have made an effort to find out about the savior. Taken aback, Liesel said that her God was certainly more compassionate. From then on, Matossian completely ignored us. We were naive enough to believe that this was an isolated case of religious fanaticism, but we now realized that such beliefs, which are particularly dangerous when used politically, are extremely widespread.

When we moved into our new flat, Arab terrorists were very active in Jerusalem and the surrounding settlements. Scarcely a night passed without an incident. Being similar in appearance to new Jewish immigrants from North Africa, Arabs found it easy to move around freely, coming from the Old City into Jewish Jerusalem.

We were witnesses to one of their acts of terror. There was a large olive grove which extended from the Jewish to the Arab part of the city near our flat. Two unarmed Jewish watchmen guarded a large building that was in the process of being put up. One night, we heard shots opposite our bedroom, fired by automatic rifles. I instructed Liesel not to turn on any lights and to crawl over to the wall facing the new building. In this way we would be protected from shots coming through the window. We took the telephone out with us and I phoned the police. The officer on duty was the husband of one of my students; he ordered us to stay where we were until he arrived in his jeep. When he arrived we turned on the search light outside our house and ran to the building opposite. We found one of the young watchmen dead, lying face down, his body riddled with shots. The other guard had also been shot, but was still alive. Liesel lifted his head carefully, and tried to calm him down. He died in her arms before the ambulance came.

Maiber, the police officer, told us to stay with the two dead men until the ambulance arrived. He believed that the terrorists had either escaped or were hiding in the olive grove. The next day, the police discovered that the terrorists had hidden in the second story of the building, after carrying out the murder and could easily have shot at us. Only after the removal of the bodies and our return to the flat did the terrorists escape through the olive grove. I had often witnessed incidents like this and acted with relative calm. But I was astonished that Liesel, who had never heard a shot fired, showed no sign of fear.

In June, 1952, the Faculty of Science and Mathematics elected me as Dean. This was a very heavy burden. Our resources were still very limited as all our equipment remained on Mount Scopus, and I had to try and deal with this situation. In this respect, I had some unexpected donations, for example funds from the Canadian WIZO (Women's International Zionist Organization). We used the money to put up wooden huts to serve as laboratories on the site where the residence of the President of Israel stands today.

On her visit to Israel, Sally Gottlieb, the President of the Canadian WIZO wanted to see "her" laboratories. Dr. Senator, the Chief Administrator, and Dr. Manka Spiegel, the Head of the Department of Organization and Information felt embarrassed about showing her the huts, thinking that she would expect to see a "real" building. I had to show her round and explained to her that we had put up huts temporarily in order to solve our immediate problems quickly and cheaply. I also used the opportunity to tell her about the long-term goal of the University to build a new campus. Would Canadian WIZO be willing to donate money for a Biology Building there? She agreed, and the first building on the Givat Ram Campus was the "Canada Hall" for science teaching.

My position as Dean often brought me into conflict with the University Administration. My confrontation with Professor Selig Brodetsky, an English mathematician, who succeeded Dr. Judah Magnes as President, was particularly stormy. At one meeting I complained bitterly about the failure to plan for the future. Brodetsky took this as a personal attack and asked me to detail my complaints. I told him the following story: "My father, Hermann Schwarz, was the owner of a large department store in Metz, which he had built up from a small shop. A neighbor of his had started in the same way, and remained a small shop-owner. Once he asked my father the secret of his success and got this reply: 'If a salesman comes and asks if I will buy some merchandise, I see what I need and then decide whether I can increase my business by selling more of those items. If so, I order much more than I actually need. This is risky, but it is the only way to get on in business. If the same man comes to you, a cautious person, you only order what you really need." Brodetsky angrily took this to mean that I thought of him as the manager of a small business. "No," I said, "You run a large business as if it were a small one."

In March 1953, Benjamin Mazar, a Professor of Archaeology, was elected Rector by the Senate. The Board of Governors also appointed him President, electing me Vice-President. As in Europe, the Rector, who is responsible for academic matters, is elected by the Senate. However, in analogy with the American custom, the President, who has control over the budget, administration and development, is nominated by the Board of Governors, consisting of scientists, businessmen and leading personalities from overseas. As both President and Rector, Mazar had unprecedented power. He faced immense problems - the dispersion of the University departments throughout the city often made it impossible for students to get to their classes in time, and with the growing student body, it was impossible to continue using the temporary buildings, barracks and huts. In addition, some of the buildings used were located in dangerous areas of the city, such as the Department of Plant Physiology, which was very near the Old City walls, where the Jordanian army guards were stationed. They had a clear view of the building from there and sometimes "practiced" shooting at civilian targets. I often had to crawl under fire to the entrance of the building.

In view of this situation, our first joint decision was to build a new centrally located campus in the Jewish part of the city to house all the University Departments. This step was approved by the Senate and Board of Governors. However, the road from decision to reality was long and torturous. How could we raise the funds necessary for building the new campus? Would the Government agree to give us the site we wanted for the campus? This brought us into unexpected conflict with both Ben Gurion, who was Prime Minister at that time and with the Hadassah Organization, which had decided to build a large new hospital in Ein Kerem, which was about ten kilometers from the center of Jewish Jerusalem. They both wanted to force us to build the University Campus in Ein Kerem too. We refused since this would involve a lot of traveling. After long and bitter arguments, the government finally agreed to allocate land to us in Givat Ram.

Opposing Ben Gurion, who was an obstinate man, was extremely difficult. He spoke at the inauguration ceremony of the Givat Ram Campus on April 27, 1958. On his way to a special podium erected for the occasion and surrounded by a wooden fence, he whispered to me: "Apparently you want to punish me and put me in a cage because I opposed you." This great man, with many more important matters on his mind never forgave us for acting against his wishes.

The hardest part in building the new campus was getting the necessary funds. The government only donated the land. The money for building had to be raised from private individuals. On June 6, Liesel and I left for an eight-month fund-raising drive in Europe and America.

Chapter 9
In Europe and North America on
Behalf of the University

Although this was not our first fund-raising trip, it was by far the most difficult. During our efforts in North and South America almost immediately after the establishment of the State, there was great interest in the new country and its University - The Hebrew University. By contrast, in 1954, the State had been in existence for six years and people tended to take it for granted, and consequently collecting money for a new university campus was much harder.

The numerous speeches to Jewish communities and societies of "Friends" in England, Scotland and Wales were the easiest part; I described the importance of the Hebrew University in Jewish life and why building a new campus was absolutely essential. Personal visits to wealthy people, from whom we expected large contributions, but who often did not consider support to the University as top priority, were much more trying. I was sometimes disgusted by these visits. Once, we were invited to the home of a very wealthy man, whose collection of jade figurines, that seemed more beautiful than anything exhibited in museums, we admired. When asked how long it had taken him to build up such an outstanding private collection, he remarked, "You mean that junk? Someone bought it for me at an auction." Fortunately, such unpleasant incidents were exceptions.

Two people in particular opened doors for us into the world of Finance, Business and Public Life: the President of the English "Friends," Norman Bentwitch, a scion of a well-known Anglo-Jewish family, who had been a member of the British Government during the Second World War, and who later became a Professor of International Relations at the Hebrew University, and his secretary, Dr. Walter Zander. The people we called on included Viscount Samuel, the first British High Commissioner in Palestine, Lord Victor Rothschild, James de Rothschild and his brother Edmond de Rothschild, who had been in charge of one of the Jewish Brigade's artillery batteries during the Second World War. Meetings were set up for us with Sir Simon Marks, Lord Israel Sieff, Alec Lerner and Harry Sacher of Marks and Spencers, the British department store chain, Lord and Lady Stansgate and Lord Nathan, who invited us to listen to debates in the House of Lords.

Shortly after I became Vice-President, Liesel and I asked Dizza Koller, wife of one of my Assistants, to organize a guided tour of the University for visitors. We also gave her express instructions to invite people who seemed lonely to lunch at our house and not to pay attention only to V.I.P. visitors.

On one tour, she came across an elderly English couple, the Kennedy Leighs, and phoned if our invitation still stood. The couple, who seemed modest and unassuming, offered to give a fund-raising party for the University in their "small London house" during our forthcoming visit to England. They then showed us photographs of their "small home," which turned out to be a large mansion in the center of town surrounded by large grounds, including a brook and a greenhouse for cultivation of orchids.

The Kennedy Leighs organized a large reception and fund-raising dinner. The morning before, one of the guests, an Australian friend of ours, called to tell us that the event had been cancelled as he just had been informed by the host's secretary, that one of Kennedy Leigh's daughters had been seriously injured in an accident. We called to check this out and were told nothing of the kind had happened and that the meeting would take place. Later, the caterer, who had brought over tables and chairs came round to take back his equipment. He explained that he had also been told that the meeting had been called off because of the "accident." Later still, all the guests waited in vain for the guest of honor, the banker Edmond de Rothschild. It transpired that he also had been told not to come by the bogus secretary. This mysterious incident was never cleared up, but we suspected a woman who was jealous of the Kennedy Leigh's social prestige had tried to sabotage the meeting. In spite of this, the occasion was a great success.

We sat next to the mayor of Greater London, who entertained us with amusing stories. After an excellent dinner, I gave my talk. Our generous host promised he would double the amount donated by his guests. We became very close, and Kennedy Leigh and his family continued to give generous support to the University, particularly to the Faculty of Agriculture. He also helped us in our work in the Negev.

We were also invited to Lord Victor Rothschild's home in Cambridge, a magnificent house dating from the 12th century. Lord Victor is a zoologist, and while we sat in the garden, his daughter Miriam, who was 11 years old at the time and is now a renowned zoologist, showed us some snails, and explained that she was studying their sexual propagation. The garden had what we came to recognize as a "perfect English lawn." We asked what we could do in Israel to create such beautiful lawns."Treat them properly for a few hundred years," he replied.

In the United States we found fund-raising on behalf of the University much more difficult than in England. Many loyal supporters of the University, including Dr. George Wise (the largest auditorium on the Givat Ram campus carries his name), Allan Bronfman (the amphitheater is dedicated to his late daughter), Sam Katzin and Prof. Wechsler, helped us in this task.

Although our main offices were in New York, during our six-month stay, we never spent more than two days in the same place. I lost count of the speeches I gave in synagogues, at business meetings, lunches, and dinners. At the end of our trip I noted in my diary: "For the first time in my life, I am totally exhausted."

One highlight of our trip was a meeting at Princeton with Albert Einstein, who was still full of energy in spite of his advancing years. Since his English was poor, we spoke German. He was deeply interested in the Hebrew University, and long before its foundation he had already stressed the importance of a University in the Jewish homeland. He gave us permission to use his name in order to raise the target sum of one million dollars for the "Einstein Institute of Physics" on the new campus. Prof. Dinur, who was the Israel Minister of Education and Culture, joined us on this occasion, speaking to Einstein in Hebrew while I acted as his interpreter.

Dinur, an old friend of mine from days before the State was founded, when he was Director of the teachers' seminary in Beth Hakerem in Jerusalem, where I taught Biology on a part-time basis, was a real gentleman. He was an excellent teacher of Jewish History and helped his students in all their problems. His knowledge of English was very limited. The day before our meeting with Einstein, he gave a talk in English at the Waldorf-Astoria on the educational problems in the newly established State to VIPs from industry and politics. Somebody had prepared an English text for him, and the evening before the lecture, he came round to practice reading it with me. He found particular difficulty with the phrase "focal point," which he pronounced as "fuckal point." Would he have said that in his lecture, the audience would have been quite amused and I advised him to replace the "dangerous" expression.

Dinur was a born storyteller. He once told us this story: Of Eastern Europe origin, as a child he was once kidnapped by gypsies.

Some time passed before his parents found him. "How do I know that I am your real son?" he asked. "Don't worry my child, God knows who you are," his father replied.

While in Florida, we visited Dr. Gottlieb, a 75-year-old retired doctor from Chicago, who was making a donation towards the establishment of the new campus. He lived alone, apart from a few servants in a palatial 14-room villa located in a tropical park with palm trees, multi-colored Bougainvillaeas and giant Ficus trees. We established a rapport with him, and he told us about his background, a typical life story of a Jewish immigrant to the United States.

Born in Lithuania, he went to a *Yeshiva* (an Orthodox Jewish school), where he was such a good student that he was ordained as a rabbi at the age of 14. A year later he started reading Hebrew books on Natural Science. Subsequently, he abandoned the faith to become an atheist. During the Russian pogroms in 1905, both of his parents were killed and he decided to go and live in America. He walked and hitch-hiked to Hamburg, and borrowed the money for the passage to New York. He did not know anybody in New York, and had no relatives in the United States. At first, he earned his living by teaching Hebrew in religious schools. Of course, he concealed his religious stance from them. Not knowing any English, he tried to learn the language from textbooks, and by listening to sermons in English in Reform and Conservative synagogues. While travelling from place to place, teaching Hebrew, he prepared himself for the university entrance examinations. He was accepted by the University of Minnesota, where he studied Natural Sciences. A brilliant student, he was then accepted into Medical School. At the age of 34, he passed his final examinations with the highest possible grades. All this he achieved without help, through hard work and unusual fortitude, driven by an inner force. He worked in his chosen profession, and later started dabbling in the stock exchange as a hobby. Through his unusual intelligence and intuition, he soon became a multi-millionaire. Eventually, he retired to Florida. Remaining a bachelor, he was a loner without real friends, and in spite of his deep knowledge of Judaism and Hebrew, which he still spoke fluently at that time, he had no connection whatsoever with the large Jewish community there. However, he was vitally interested in the fledgling State of Israel, which he hoped would be "Jewish," rather than being based on a religious framework.

During our conversation, his telephone rang incessantly - a stock broker from New York, then one from Chicago, and so on. I was unable to restrain myself from asking him why he continued to play the stock market after having suffered a serious heart attack. After some hesitation he explained that he had no interest at all in money, but was "addicted" to this activity. We felt sorry for this rich lonely old man, who was incapable of really enjoying life.

I was scheduled to give a talk shortly before leaving for Detroit. I asked Liesel to pack in the meantime, and we planned to meet at the airport at 10 p.m. However she felt so ill that the chambermaid had to pack for her. When I met her at the airport I saw that she was terribly sick. It was New Year's Eve, and thoroughly exhausted, we stopped over in New York. At 4 a.m. we got a call informing us that all flights had been cancelled due to snow. We decided to travel by rail. The train was full, and only after giving the conductor a New Year's "present," did we get two places in the Club Car. We traveled through the night, and Liesel complained of pains in her back. After arriving in Detroit we called a doctor immediately.

Liesel had viral pneumonia in both lungs. According to our schedule, I had to be in Boston the following day to present a talk to the Jewish community and to lecture at Harvard and Brandeis Universities. I left Liesel, who was justifiably angry, in the care of my cousin Freda Krohn. With hindsight, I think that I took my duties far too seriously at that time.

During my first talk in Boston, I suffered from acute abdominal pains. On the return journey to Detroit, I fainted in the dining car, but before passing out, had the presence of mind to write Liesel's address in Detroit on the table cloth. The conductor immediately radioed the next station to get hold of a doctor. Dr. Rubin examined me there, and concluded that I did not have appendicitis or a heart attack. He gave me an injection of pain killer, instructing me to call a doctor the moment I arrived in Detroit. He was extremely nice and calmed me down - as the pains and fainting had frightened me. On realizing that I came from Israel, as a Jew and a Zionist, he refused to accept payment for his services.

In Detroit, doubled up in pain, the nice train conductor carried me to a taxi. I made my way to the hotel room, where Liesel was still in bed with pneumonia. Instead of being able to help her, I meekly crept into bed next to her. The doctor told me that I must undergo an immediate operation for diverticulitis, but this time I refused to leave Liesel. The doctor could not find a hospital that would admit both of us, so we remained in the hotel room. This proved to be a stroke of luck since the pains miraculously disappeared the next day! We stayed in bed for several days, sustained on a diet of chicken soup, the well-known panacea of Jewish mothers, brought to us daily by Freda.

Two days before our flight to Europe, during my last lecture in Washington DC, talking to a group of VIPs, including senators and congressmen, I felt terrible pains in my back, and only managed to remain standing by clutching the rostrum with all the force I could muster. Nobody seemed to notice that I was having some difficulty in speaking. After the lecture, a doctor told me that some of my discs had been dislocated.

I was carried to the plane on a stretcher and placed on a bed in the first-class compartment - the first and last time I flew first class. In Zurich I received injections of pain killers, so that I could travel on to Davos, where we were going to spend vacation with Liesel's aunt.

After I recovered somewhat, I went skiing, a sport I had last practiced in 1936. The moment I put my skis on the soft springy snow, the pains in my back seemed to disappear. Our Swiss instructor Stucki, who had no knowledge of English, and therefore had difficulty in communicating with this class, came out with many funny statements: "Put ze back in ze front and go in the neuse (knees)." I was his worst student since I had completely forgotten how to ski and could not manage the difficult turns.

On the last day of the course, without any conscious effort on my part, my knowledge of skiing seemed to come back, and I gave a perfect performance. Stucki could not believe his eyes, and asked me to repeat the run, which I did perfectly. Rather angry, he declared in his best Swiss German, "The damned chap has been fooling me all the time. All these professors are mad. I am going to shoot him with a pistol." After explaining what had happened, we had a good laugh, and went to the bar to celebrate.

As in the past, during this fund-raising trip, I also managed to visit scientific colleagues and institutions. I also participated in the Eighth International Botanical Congress in Paris, which took place in the old Sorbonne buildings. It must have been the most badly organized congress ever held - there were no elevators, so that in order to find the lecture halls, we had to walk from floor to floor. On one occasion, I had to search the whole building for a slide projector for the lectures I had organized, which began 20 minutes late as a result. An Indian colleague was due to give a lecture on the effect of music on plant growth, but his "audience" found that the lecturer did not turn up and the hall was locked.

During the congress, I also met, for the first time since 1933, some of my former German colleagues. Once, I suddenly found myself face to face with my former boss, Ernst Pringsheim of Prague, who had since gone to live in England. At first, he was very formal, but gradually became more friendly. During one of my lectures, I was overjoyed to see Bruno Huber, my former boss from Darmstadt, in the auditorium. After the lecture, we embraced each other like old friends, a gesture quite unusual for German professors. This pious Catholic, anti-Nazi, told us what he had been doing since 1933. During our contacts with this good man with kindness radiating from his blue eyes and who had done so much for me, Liesel lost some of her aversion toward Germany, and for the first time since leaving that country, she spoke to a German in his native tongue.

My meetings with other colleagues I had known in Germany, and who were Nazis or at least sympathizers, were unpleasant. They seemed embarrassed to see me, and told me how much they had suffered during the war and how they had not been Nazis. Listening to them, one might have thought that the Nazi period was a mere figment of the imagination!

After the congress, Liesel and I visited the places where we spent our childhood. With the exception of my very short visit to Metz as a soldier with the British Army, I had not seen my home town for 36 years. The visit touched me more deeply than I had expected, as I noted in my diary:

"Strolling through the town, long forgotten youth is revived. Wandering on clouds through a screen of dreams, doors that I had unconsciously sealed seem to open up. The school and the sweet shop opposite are still there. My parents' department store Hermann Schwarz is now the "Louvre," with Mademoiselle Lux still serving at the same counter as in 1918; and the synagogue, which was used by the Germans as a food depot from 1940 to 1945, still stands. I have changed so much, while Metz, after going through the War, the Holocaust, occupation by the Germans, and then being retaken by the French, remains the same. This confrontation with my past, which I thought I had eradicated from my memory, liberated me in a strange way, as if a light has suddenly illuminated the blacked-out past. I was overcome by a feeling of peace."

Something similar happened to me when I visited my father's grave in Frankfurt. I met my friends, Hans and Ille Petersen and Eva Steinschneider, with whom I had lost all contact since 1933. The Petersens, who were Quakers and anti-Nazis, had been imprisoned for distributing anti-Nazi pamphlets and helping Jews to leave Germany illegally. Eva, a communist, had survived in the French underground, but her husband had been caught, tortured and then murdered by the Gestapo.

I also visited Kiedrich, a small village on the banks of the Rhine, where my sister Erna and my brother-in-law Gerson had lived, as described in another excerpt from my diary:

> "I visited Anton Krams, the sculptor, who was like a son to Gerson. He is still angry with them for leaving the "Vaterland" (fatherland), and clings to his naive belief that the Catholics of Kiedrich could have saved the Sterns from the Nazis. I tried to explain the situation to him. During our discussion of Hitler's regime, I mentioned the word "Gestapo." His daughter asked me what that was and told me that when students mention the Nazi period in school, the teachers evade the topic. This seems to me to be a clear attempt to blot out the disagreeable past on the part of the teachers, who had probably been Nazis. I came across similar attitudes in other places in Germany. Although psychologically understandable, this dangerous refusal to come to terms with their dreadful past bodes ill for the future."

Liesel went to Bitburg, where her grandparents had brought her up. Her mother had died while giving birth to her, and her father had been a soldier in the German Army. Bitburg was Liesel's real "home" although she had been born in Saarlouis (Saarland). We went from grave to grave - our only ties with our old homeland. Her mother is buried in Saarlouis, her grandmother in Luxembourg, and various ancestors in Bitburg, where the family lived for hundreds of years. Jews from all over the world - Marocco, Libya, Algeria, Yemen, Hadramauth, Afghanistan, China, and India - go through similar experiences when they visit the graves of their families and famous rabbis.

During the Battle of the Bulge in 1944, most of the town had been destroyed, but the house belonging to Liesel's grandparents, and the hotel nearby, where we stayed, remained intact. The Jewish cemetery had been laid waste by the Nazis, and the gravestones had disappeared. During the War, the Germans used the jewish cemetery for burying Russian prisoners-of-war, whose bodies were subsequently exhumed in 1948. In the garden of one of the houses there we found the gravestone of Liesel's great-grandfather, a large black obelisk which was later re-erected in the Jewish cemetery. This was all that remained of what had once been a prosperous Jewish community, and was later to stand alone in the large Jewish burial ground.

As well as defacing the Jewish cemetery, the Nazis had also erased the name of Liesel's uncle, a German officer, who had been killed in the First World War, from the memorial to fallen soldiers. After the Second World War his name was re-installed.

Mr. Simon, the owner of the well-known brewery in Bitburg, recounted how people of Bitburg had refused to burn down the synagogue on the Crystal Night in 1938, and when Nazis from Kylburg came to burn it down, they had been driven away. The synagogue was saved by using it as a store for wheat. Unfortunately, the old synagogue building was destroyed later during the Battle of the Bulge. Liesel was heartbroken at this since the synagogue had been built by her ancestors. With a clearer picture of the significance of our past, we returned to Israel to confront the problems of the future.

Chapter 10
Vice-President of the Hebrew University

During my term as Vice-President of the Hebrew University, I did not give up research and teaching. This proved difficult, as together with Mazar, who was the President as well as being charged with routine administration of the University, we were assigned the task of getting the new campus on Givat Ram built. Much of my time was taken up with this. Every building was planned by a different architect and they wanted to erect monuments showing their geniality that did not generally take into account the primary function of the buildings. After much discussion, we succeeded to some extent in adapting these plans to suit the prevailing conditions trying to steer clear of unnecessary luxuries such as massive entrance halls, expensive marble tiles and mosaic-covered pillars.

The building supervisor, an able but very independent person, often acting on his own and without taking into account financial constraints, was also difficult to work with. The new campus site was in a wild, rock-covered area with irregular clefts and fissures. We planned to develop the land between the buildings into a green area with many trees. Mazar and I had decided to leave one large piece of rock at the entrance to the campus showing the original state of the site. After an argument, the supervisor agreed to this but one morning the rock mysteriously disappeared. The supervisor explained that his workmen had removed the rock contrary to orders!

I also had problems with my colleagues in the Botany Department, who claimed that I neglected their interests. Although Canada Hall was nearing completion, together with two one-story buildings which seemed to me to be ideally suited for use as lecture hall and laboratories, the entire science department refused to move into the buildings since the campus was some distance from the city, and there was still no regular bus service. I asked my own department to move into the buildings as pioneers, a request to which they reluctantly agreed. However, within a short time, building after building went up and the bus service functioned well, and I was then accused of having favored the Botany Department!

The administrative system of the University was also irksome. Building a new campus in a young State for a University beset with innumerable problems including financial limitations, through a complicated and cumbersome organization which we had inherited from the days of the Mount Scopus Campus, was no easy task. The Board of Governors, the highest authority in the University, consisted of between 50 and 100 members, who met biannually; the Standing Committee with 15 Israeli members and 15 governors from overseas and the President, Vice-President and Rector as ex-officio members, convened biannually in years when the Board of Governors did not meet. An "Interim Committee," a subcommittee of the Standing Committee, which included the Chairman and Vice-Chairman of the Board of Governors, President, Vice-President, and Rector, was responsible for making final decisions concerning hiring and promotion of tenured academic personnel. The Executive Council, a representative body of the Board of Governors, included the Rector, Pro-Rector, six professors elected by the Senate, six government representatives and 14 members elected by the Board of Governors. The Permanent Committee, responsible for the day-to-day administration of the University had 25 members, including the President, and the three highest-ranking members of the administrative staff, including the Vice-President, and two representatives of the academic staff on the Executive Council, including the Rector. Last, but not least, the Senate, a purely academic institution, had the power to intervene in University affairs.

It requires little imagination to appreciate how difficult it was to work with all these bodies. Hardly a day went past without having to go to meetings of one of these forums with their endless discussions. Most of the decisions taken could have been made in a fraction of the time, and I often could not conceal my impatience with this waste of valuable time. In addition, we had to raise all the funds for our building projects, and consequently Liesel and I continued our fund-raising efforts until 1959.

I could never have endured the seven years I spent as Vice-President without the satisfaction I gained from certain aspects of this work - the feeling of helping to build something new, both in the physical sense on the new campus, and in the introduction of new academic fields.

In keeping up with many American Universities we decided to introduce a school for Home Economics into the Faculty of Agriculture. This proposal had to be submitted to the Senate for approval. The discussions on this matter stand out in my mind; learned scholars, including Gershon Sholem, the well-known researcher of

Jewish *Kabbalah* and mysticism, were violently opposed to such a "school for cooks" because it did not exist in any European University. It took a long time to explain to them that in the American system, rather than the European one, Home Economics was a scientific branch in the discipline of Agriculture. We stressed its importance in Israel, a melting pot of immigrants from many countries and cultures. In the final vote, a majority voted for the establishment of this new school.

We had similar troubles when we proposed the establishment of a School for Business Administration within the framework of our Faculty of Social Sciences. George Wise, the Chairman of the Board of Governors originally came up with this idea as most American Universities had such schools. Following the customary fights in the Senate, a majority voted for the new School. We then invited the Dean of the School of Business Administration of the North-Western University in Chicago to Jerusalem to advise us on how to go about setting up this new institute. Mazar asked me to lead the negotiations. Following discussions with the professors in the Faculty of Social Sciences, the Dean feared that the School would be biased toward theoretical rather than the practical aspects. He only offered to help us on the condition that I participate in one of the summer courses in Business Administration at his Institute, since he was convinced that I would keep the school on a businesslike level.

As a result, for a six-week period in 1956, I studied Business Administration in Chicago! There, Liesel and I were invited to dinner by the Dean and his wife. The Dean's wife, who had accompanied her husband on his visit to Israel, remarked, "The Israelis are the most wonderful people I have ever met," and turning to her husband, continued, "Isn't it true that we don't know any Jews besides the kikes in the New York subway?" Liesel immediately retorted, "Isn't it wonderful? A short time ago we were also kikes." I did not dare to look the Dean in the face. This exchange gave us some inkling of how different these circles were from those I had come across on previous visits to America.

The participants in the summer course lived in the building of the School of Business Administration and were allowed to leave the University only on weekends. On the first day I already felt like a black sheep. After registration, we all went down to a bar, which had been specially installed for our course, and subsequently met there every day at six in the evening for the so-called "happy hour." From the start we addressed each other by our first names - I was "Mike." My colleagues were all Vice-Presidents of large American business enterprises, such as Bethlehem Steel and General Motors. They had been sent on the course by their companies in order to learn about new methods of business administration. After a few whiskies, they began asking one another about their respective functions in the corporate world and then what they were "worth." I was taken aback, and could not visualize similar questions being asked in European circles. Their monthly salaries ranged between $20,000 and $40,000, without taking into account fringe benefits, such as the use of a chauffeur-driven car and living accommodation. I had to admit that my monthly pay was less than 5% of their minimum earnings and said: "I hope I am 'worth' more!"

I was the only Jew. They found out that I was an Israeli, but were not sure whether this meant that I was Jewish. During the "happy hour," they cautiously began questioning me about Israel, the people who lived there - obviously, since I was not an Arab, I must be a Jew. They used the word hesitantly, and only later did the reasons for this become clear to me. On confirming that I was Jewish, one of them said that I must come from a different "tribe" than the Jewish businessmen with whom he dealt. He went on to explain that the very mention of the word "Jew" by a non-Jew to the American Jewish customers he came across would offend them. By openly stating that I was Jewish, I had broken the ice, and we had a lively discussion about Jews and Judaism.

These people had strange ideas, often anti-semitic, about Jews. None of them had ever had any "real" Jewish friends. Having come into contact with me, they wanted to know all about our history, religion and customs. They invited me to give lectures about Israel, Jews and related topics in the evenings. Hopefully, after my lectures and our discussions, they lost some of their prejudices, which were not limited to Jews, but also covered other minority groups.

Politically, they were all ultra-conservative, and I had the impression that they had not done any serious reading since leaving college or University. During the course, we were required to read various texts on economics such as J.K. Galbraith's writings (28, 29, 30). My fellow students were shocked by Galbraith's "revolutionary" and "socialist" ideas, and his criticisms of American society. Discussions often became violent when they realized that their own ideas were totally opposed to those expressed in his books.

During one of these discussions, we talked about poverty. Why did poverty exist in modern, prosperous America? Most of them considered that the poor were simply too lazy to work, and that rather than trying to help these "parasites" one must do everything possible to get rid of them. Knowing that I was somewhat of a maverick and non-conformist, Ch. J. Gragg, whom we nicknamed "Chuck", who was a genius as an instructor, winked at me. He asked for my opinions on poverty. On the verge of exploding, I tried to answer as calmly as I could, asking them whether they had heard of people who believed that possession and wealth were sinful. Had they forgotten the words of their own Jesus: "It is easier for the camel to go through the eye of a needle than for a rich man to enter the Kingdom of God" (Matthew 19:24). Had they never heard of Francis of Assissi (1181-1226), the most revered saint of the Church, who renounced his family richness and preached the blessings of poverty? After a few minutes of silence the discussion continued with a different tone, a radical change from what they had said earlier.

At the end of the course, we got our diplomas and were no longer strangers. We had mutual contact with worlds previously unknown to us. My colleagues did not forget me and for many years to come they sent me Christmas cards.

Besides teaching me the basic principals of Economics and Sociology, I also learnt a new teaching technique - the Case method. In place of lectures, we had to propose solutions to real cases. For example, we were told that a certain corporation was on the brink of bankruptcy, and given details of its administrative structure and financial situation, including balance sheets of income and expenses, and data concerning the size and structure of its work force. Divided into two groups, in the evenings, we would work out a plan to save the company and make it a going concern, appointing members of the group as President, Vice-Presidents and Chairman of the Board. It was the only time that I was "President" of a manufacturing company employing 10,000 people. The following day, each group put forward a proposal for saving the company and we discussed these. The teacher would then point out the good and bad aspects of the two decisions, until we reached a mutually agreeable solution. I later applied this method successfully in teaching, and found it more efficient than lectures because the students had to work out solutions for their problems by themselves.

On my return to the Hebrew University, I resumed my duties as Vice-President. One of the most interesting aspects concerned the reception of official University guests, whom we invited to meals in our home. We met outstanding scientists including Niels Bohr and other

Nobel Prize winners, famous writers, leading diplomats, politicians and Government Ministers.

With the exception of Ben Gurion, Golda Meir was the most interesting of the Government Ministers we entertained. In her honor, we invited the most prominent professors, who were quite bashful and somewhat overshadowed by her powerful personality. It was said of her, that she ruled with an iron hand. My confrontation with her on one occasion bore this out. I went to see her on University business, and after listening to my request, she simply said "No." Refusing to discuss the matter further, she reprimanded me and stood up in order to indicate that our meeting was over. I left the room, feeling like a schoolboy, who had just been ticked off by a teacher, in this case by "the only *man* in the cabinet," as people called her.

We also got to know a completely different and unexpected aspect of Golda's character. As Foreign Minister, she once invited us to her home together with some of her other friends. In this intimate circle, she told us how, disguised as an Arab woman, she had gone to meet Emir Abdullah of Jordan in his winter palace. On hearing this fascinating story, Liesel told Golda that she should write her memoirs, to which Golda answered, without a trace of false modesty, that it was for more important people than her to do this. She later wrote an autobiography, but only after she had served as Prime Minister during one of the most critical periods in the history of Israel - the Yom Kippur War of 1973. Typically she did not try to gloss over some of the serious mistakes she made during her long political career.

The ambassador of the Soviet Union was a guest in our house before the Soviets broke off diplomatic relations with Israel. Before lunch I showed him round the Faculty of Medicine, temporarily housed in a building in town, previously used by the British Bible Society. That day, the students were on strike.

Noting the primitive conditions there, the Ambassador tactlessly described the excellent medical facilities in his country. Later, during lunch, looking at the professors seated round the table, he provocatively remarked that in the Soviet Union such parties always included students. I replied, that this would not have been possible today in any case as the students were on strike, and asked ironically whether the students in his country had the right to strike over grievances against the administration. He chose not to answer.

In another "diplomatic" encounter I met Adlai Stevenson in 1952, after his defeat in his campaign for the Presidency against Eisenhower. He came to Israel after a visit to Jordan. As Teddy Kollek points out in his autobiography (31): "When Stevenson came to Israel, everything seemed to go wrong. I, for example, accompanied him to the provisional building housing the administration of the Hebrew University in order to meet Prof. Mazar. From the window, we could see the border at Mount Scopus, from which we had been cut off since the War of Independence. Mazar explained that at Mount Scopus was our national library, our books, and our land that was all inaccessible to us." Stevenson, who had visited some refugee camps in Jordan that morning, pointed out acidly that the Arabs had the same feelings when they looked from their side to ours. There are several inaccuracies in Kollek's account - Mazar was not accompanying Stevenson, I was! Also, taken aback by Stevenson's remark, Teddy in his usual outspoken way whispered to me in Hebrew, "You damned idiot, how could you say such a stupid thing to our guest?"

In an interesting follow-up, two years later while on a fund-raising trip in Chicago, we went to a rather unpleasant celebration honoring an important Jewish philanthropist, who was also involved in American politics. Recognizing me, Adlai Stevenson, who was present together with other senators, greeted me like an old friend and said in his outspoken manner, "Professor Evenari, thank God for at least one human being!" To my great surprise, he told me how much he had enjoyed his visit to Israel.

We once entertained the Comte de Paris, the Bourbon pretender to the French Crown. Showing us photographs of his eldest son, he declared that this boy would be King one day. Liesel mistakenly took this as a joke and laughed. A perfect gentleman, the Comte ignored her remark. He invited us to visit him in Paris, and when in Paris we took him up on this offer.

Another time, a mysterious commission of high-ranking French officers lunched in our house. Mazar asked me to keep their visit secret. The group was led by an Admiral, and all his colleagues were generals in the French army. They politely refused to write down their names in our guest book, as was customary. They did not reveal their names and so we addressed them by their titles - "Monsieur le general" or "Monsieur l'amiral". I could only guess why they were in Israel! Our lunch together was a near disaster. Until then, Liesel had always employed a cook to prepare lunches and dinners as we often entertained from twelve to twenty people at one of these meals. On the day of their visit, we had a new cook, Christina from Yugoslavia. She seemed to be a good cook, but was rather excitable and nervous. Liesel went into the kitchen, and asked her to put the chairs around the table, at which she went into a huff and walked out slamming the door. Initially taken aback, Liesel then called in a waitress and finished preparing the lunch herself. In the meantime, our French guests had arrived, and I drew out the aperitifs until I got the all-clear sign from Liesel that lunch was ready. After the meal, Liesel told the Admiral what had happened. In the typical manner of French aristocrats, he kissed her hand and said, "Madame, I admire your sang froid. Many other women would have panicked under similar circumstances, and the guests would have noticed this. We didn't. From then on, Liesel always did all her own cooking for these luncheons.

We once entertained Lord and Lady Stansgate, who had just returned from an official visit to the neighboring Arab states. They had seen many Palestinian refugee camps and were horrified by the terrible conditions under which the people there lived. In Syria, their official guide had shown them vast fertile areas that had not been put to agricultural use. Lady Stansgate asked her guide why these regions were not cultivated. He explained that the population of the country was so small that there were not enough people to settle the areas. She naively suggested that the Palestinian refugees would be more than pleased to go and live there. In reply, the guide said that there was no need to settle this Palestinian "rabble" in their country. Regrettably, this attitude has not changed since then.

Lord Hore Belisha and his young wife were interesting guests. Belisha, who was Jewish, had been British Minister of Defense at the beginning of the Second World War. Since it was their first visit to Israel, we asked him how he liked the country. "I was a fully assimilated Jew and thought that I was an Englishman just like everybody else in England. But when my plane landed in this country, I felt that I had come home." At this remark his wife, who was non-Jewish looked at him with a stony expression.

My son Eli often came to these lunches. Once we were entertaining a Professor of Philosophy from Cambridge who spoke perfect "King's English." Suddenly Eli said, "I don't understand one word of his English. He speaks with such a terrible accent." Initially speechless, the professor burst out laughing.

In 1956, while on a fund-raising trip, by chance, I came across a British officer I had served with during the War strolling down Oxford Street. This passer-by embraced me - a quite unusual event in England. With tears in his eyes he shouted, "Sergeant Major Evenari! Sergeant Major Evenari!" It was Captain Johnson, who was the most well-liked officer of I.A.A. Battery, who was now working as a bank clerk in London. We talked of old times, and he remarked several times how "The period I spent with you and your people was the best time of my life."

Another time, Liesel and I were in a bus going along Fifth Avenue. Through the window, I saw a tall man, whom I immediately recognized as Major Rosenberg, one of the former commanders of our artillery unit. I told the bus driver to stop and let us off. Running after the man, I shouted out his name. Startled, he turned round, and seemed delighted to meet me after so many years. He asked me how I could possibly have recognized him from behind and from a moving bus after so long. I felt embarrassed to tell him that I recognized him because he had the most prominent flat feet I had ever come across and waddled like a duck!

Later that year, on October 29, we were in London. Liesel listened to the radio and suddenly shouted to me that Israel, England and France were at war with Egypt. We decided to return to Israel, but all flights to the Middle East had ceased. I phoned our embassy to see if they could help us return home. I was told that at my age the war could be fought without my help. So we stayed in London and kept to our schedule.

We had already received an invitation to visit the Parliament for October 31. We went, and were witnesses to an unusual event. We met Lord Hore Belisha, Lord Nathan, Lord Silkin, Lady Reading and Viscount Samuel who were as surprised by the war and the strange alliance of Israel, Britain and France as we had been. They asked me if I could explain why Israel had gone to war with Egypt. I said that as a non-politician I didn't know, but that it was clear to me that sooner or later, a confrontation between Israel and Egypt was inevitable. As I knew from experience, the almost daily terrorist attacks coming from Egyptian territory were insupportable.

We then attended sittings in both Houses during the debate on the war. We were astonished by the Zionist speech of Anthony Eden, who was then Prime Minister, and who was not known for his Zionist sympathies. Only then did we understand the British and their attack on Egypt. I will never forget the beginning of Lord Silkin's speech: "Today I am speaking to you, my Lords, as a Jew, not as an Englishman."

During our fund-raising trips I tried to attend scientific meetings, to keep in contact with my colleagues, in the countries we visited. The most important ones were a number of symposia of the IBP (International Biological Program), organized by UNESCO on Arid

Zone research. The meetings took place in Paris, Montpellier and Madrid. These symposia were the only opportunity I had to meet my colleagues from Arab countries hostile to Israel. Carefully avoiding political discussions, I became good friends with Professor Mohammed Kassas from Cairo University. Other Arabs, however, refused to talk to me. Kassas, Rodin and Kovda, who came from Russia, and I were in total agreement on most issues and defied the majority view.

In a discussion on the Bedouins, I was the only one to speak up for them. Everyone else thought that they should be forced to abandon their nomadic life-style. This attitude annoyed me; I declared that we must do all we could in order to preserve their social structure. There were ways to prevent overgrazing of deserts and semideserts, without forcing them to become sedentary. Ironically, the only Israeli in the group, I stood up for the Bedouins against my Arab colleagues. Later, my fight for the Bedouin was unsuccessful even in my own country.

At another symposium in Madrid, I met Otto Stocker, who had served as my paragon for Desert Research since my early student days. As I had not received the final program since I was away on a fund-raising trip, while in Darmstadt, I dropped in to see Stocker, who was Director of the Department of Botany, to ask him about details of the program of the symposium to which he was also invited. In a reticent and reserved manner, he gave me the information I needed. His cool reception annoyed me and the ice between us was only broken later during our excursions together in Spain. We became close friends, and he visited us nine times in Israel and stayed with us both in Jerusalem and in the Negev. During his last visit, although 85 years old, he still managed to go alone on long walks in the desert, climbing the mountains like a young man.

We nicknamed Stocker Kaiser (Emperor) Otto. Liesel comes from the well-known Kalonymos family. Jewish historians, including the renowned Simon Dubnow (32), relate that a Jew named Kalonymos served as a retainer to Otto II and fought with him against the Saracens at Cotrone in Italy in 982 AD. Otto II owed his life to Kalonymos, who gave him his horse after the Emperor's horse had been killed by the enemies. Otto II asked Kalonymos how he could reward him. The Jew requested permission to establish religious seminaries for Jewish studies in Germany, where some of the Jewish communities dated back to Roman times. The family established religious colleges in Mainz, Worms and Speyer; another branch went to Narbonne, where Kalonymos Bar (son of) Todros was appointed

Head of the Jewish community by the King of Provence. I still have his coat-of-arms on which Kalonymos is referred to as "King of the Jews." Many members of this family were famous writers. The name Kalonymos features in most textbooks on Jewish history. After Stocker heard this story, he started addressing Liesel as "Princess Kalonymos," and in turn, we called him "Kaiser Otto."

In all respects, Stocker was an outstanding man. His knowledge, based on personal experience in the deserts of Egypt, Mauritania and the Sahara, helped me immeasurably with my own desert work. He was one of the few botanists with whom I came into contact with whom I could discuss the cognitive foundations of science, as well as other aspects of Philosophy. He went into modern Philosophy and Epistemology in great depth. He was in full possession of his mental and spiritual capabilities beyond the age of 90. With his death shortly before his 91th birthday, Botany lost a pioneer in Desert Ecology, and I personally lost a great friend.

Kassas also participated in the symposium held in Spain. While in Murcia during the Jewish New Year (*Rosh Hashana*), Liesel told Kassas during lunch that it was customary to eat on this day apple and honey, symbolizing the wish for a sweet year. Kassas responded by bringing an apple to the table, cut it into halves, giving one to Liesel and eating the other half himself. Pointing at Liesel he said, "Now I am eating apple with honey."

In spite of my activities as Vice-President, I managed to continue my experimental work. In retrospect, I do not know how I found the strength to do this. I was working between 12 to 14 hours a day. During this time, the Plant Physiology Section, which I headed, directed its efforts toward the Physiology of seed germination. Together with my assistants we published 103 scientific papers on this topic between 1952 and 1960. We were known internationally as "The Jerusalem School of Germination Physiology" among our colleagues overseas. Team-work, involving a multi-faceted approach proved most fruitful. My co-workers, Alexandra Poljakoff-Mayber, Dov Koller, Alfred Mayer, Shimon Klein and Moshe Negbi, all became professors in their own right. Today, they each pursue independent lines of scientific research, and our collaborative efforts in Germination Physiology have become part of the history of the discipline.

Chapter 11
The Negev

In the Spring of 1954, just before undertaking another lengthy fund-raising trip to Europe and the United States, Dov Koller, my Ph.D student and assistant, took me to see the gigantic ancient agriculture system that he had just visited in the Negev. Together with his wife Dizza, we drove to Wadi Ramliyeh near the ancient town of Avdat in the Negev desert. Many old terraces, huge stone walls, channels and fields covered an entire arm of the Wadi. Suddenly, the memory of what I had seen some years earlier flooded back. I was in a trance-like state and decided that the time had come to try and find out how agriculture could have flourished in the desert.

Back in Jerusalem, I searched for books on ancient desert agriculture. To my amazement, I found that I had the classic in this field, written by Palmer and published in 1871 (33) in my own library. My brother-in-law Gerson Stern had given it to me as a parting present when I left Germany in 1933. Could this have been a mere coincidence?

It soon became clear that I would not be able to investigate this problem alone. Koller introduced me to Naftali Tadmor, nicknamed "Kofish," a student at the Hebrew University's school of Agriculture at the time, who was as enthusiastic as I was about the Negev. In spite of the differences in age and temperament, we immediately felt drawn to each other. Kofish introduced me to Leslie Shanan, a friend of his, who was a hydrologist and water engineer. Shanan, originally from South Africa, became the third member of our team. We referred to the late Johanan Aharoni for help on archaeological issues. We worked together for 20 years until Kofish's untimely death.

During the next 4 years, whenever I could get away from my administrative and fund-raising duties, we went down to the Negev. Traveling in jeep or flying over the area, we surveyed and mapped close to a hundred ancient agricultural systems. With the cooperation of the Air Force, we used small Piper Cubs, which hold a pilot and one passenger. Flying low, one could open one side of the plane. Tied to a belt one could lean out to study or photograph items of possible significance. From the air, we observed many more details we would have missed from the ground. I loved the sensation of "real flying," floating freely over the ground, an experience which one does not have in the large, modern planes.

These expeditions to the Negev were difficult since most of the roads did not exist at that time. We could only travel over rough terrain in jeeps, and had to carry guns in case we encountered Arab marauders. On army orders, we had to leave the Negev before dark and stay in Beer Sheva overnight. The logistics fell upon Liesel. She had to see that there was enough food and prepare meals for the research team and volunteers, up to ten people in all. Her "desert salad," into which she put anything that was edible, became well-known. One guest said jokingly that she may have even put his sandals in one of her concoctions!

During our Negev excursion we were sometimes in physical danger as well. Once we were surveying Wadi Abiad, near Shivta, where there were extensive traces of ancient agriculture. Before starting out on this trip we had to inform the military authorities as they often carried out training exercises in this area and ask for permission to work there. While surveying the area, shells started exploding all round us. We later found out that the army command had forgotten to inform the battery exercising there of our presence in the area.

In retrospect, it seems remarkable how we worked together harmoniously for so long, although there were problems. Kofish was a rugged individualist, who had absurd but often brilliant ideas. He possessed unusual strength which the rest of us could not match. Leslie and I would often be dead tired, while Kofish as fresh as ever would force his "slaves" to work on. Once while loading up his jeep he strained his back. Liesel gave him some aspirins to alleviate the pain, telling him to take two immediately and not more than eight per day. "If so I will take sixteen!" he retorted. His powers of observation and sense of direction were uncanny. From the air, I often saw some small structures in areas we had not yet surveyed. Kofish would study my photographs briefly, and then, as if sleep-walking, drive us directly to the point we wanted to study. He could drive like a demon!

We once spotted a Nabatean fortress on the top of a very steep hill near Shivta, and, since we reckoned that it would be impossible to drive there we started climbing the hill. Kofish pointed out that it would take us at least an hour, and that he could drive us up in a few minutes. He then raced up the rock-covered slope, and we arrived unhurt, with the jeep intact. His willingness to help others also knew no bounds. I lost count of the number of times he relieved us of the heavy loads we were carrying on our backs although he was already loaded down.

In contrast, Leslie Shanan is a rather cool personality, not given to impulsiveness, with a critical mind. He was the only one who had the talent and the knowledge to draw detailed maps of the areas surveyed.

From the start, it was clear that we were confronted with a far more complex problem than we had anticipated. Where did the desert farmers get their water? Was the Negev a desert in ancient times as today, or did it have a more humid climate with more rainfall? Was agriculture restricted to a certain type of soil? Did they employ different farming methods? Were the Nabateans the only group to carry out farming activities in the desert?

When we began our work, the meteorology, climatology, soil conditions and geology of the Negev were very poorly documented. We were fortunate in that several specialists in these fields also began their investigations at around the same time as we did. Their data helped a great deal. Already from our first excursion, we realized that flash floods caused by the runoff of rain water rather than direct rainfall, served as the water source for ancient agriculture. Only much later, however, did we come to understand the mechanisms behind these phenomena.

Early on, we discovered that flood water agriculture in the Negev was limited to the Negev Highlands at heights ranging between 300 and 1000 meters above sea level, which had a mean annual rainfall of up to 150 mm. In the *Aravah* plain there had clearly also been ancient agriculture but of a different type. While flying over an area near Kibbutz *Yotvatah*, Kofish spotted "chains" of regularly spaced round holes resembling bomb craters extending for about four kilometers. On a second flight over this area, I observed long, straight lines that looked like channels near these "chains." These lines led into a flat "chessboard" area, consisting of regularly - spaced holes. It seemed to me that the "chains," "channels" and "chessboard" were part of an organized agricultural system.

On return to Jerusalem, I looked for references to such "shell holes" in the literature and found that a German researcher Frank and an American investigator Glück had described them, but had apparently not understood their function. However in a paper published many years ago in an obscure Israeli journal, Aisenstein, an engineer, had correctly referred to these as "kanats," and this also led me to other articles on the subject dealing with their function and history. Invented by the Persians around 800 BCE, kanats carried underground water over long distances to agricultural areas or as drinking water to towns. I will later explain how they functioned.

Terraced wadis were the first ancient runoff agriculture system we studied in depth. From the air, we could see large numbers of small wadis resembling staircases. On closer observation, we found out that these consisted of steps, which were low stone walls. The terraces between them contained soil. We discovered that flood waters would cascade down these wadis. The runoff water streamed from the hillside down the wadis, each terrace wall retaining a small amount of water, infiltrating the terrace soil and thereby enabling the cultivation of crops or pasture plants. The overflow water then flowed down to the next terrace. Only the small secondary or tertiary wadis were terraced. After seeing floods in large, primary wadis, it became clear why this was so: In the large primary wadis the floods were torrential and terrace walls could not have possibly resisted the mechanical force of the streaming water. This also explained why these wadis were covered with stones of various sizes, and contained no soil. We found that the terraced wadis were terraced from their sources on the hillside to the point where they merged with larger wadis, since partial terracing would have exposed them to erosion. Thus, the terraced walls had two functions: to retain water, and to protect the wadi against erosion by controlling the mechanical force of the flood water. The fact that after two millennia most of these terraced walls are still intact underscores the latter point. In some of these wadis repaired by children from the kibbutz of *Sde Boker,* onions and other annual crops were successfully grown.

Terraced wadis are the most primitive system of flood water farming. A more complex method, which we called the "farm system," near the ruins of the cities of Avdat, Shivta and Nitzana, proved more difficult to understand. Flood plains and flat wadis were covered by fields containing five to twelve terraces surrounded by a solid stone wall. Each farm system probably belonged to one peasant family as the remains of houses in many of these "farms" showed. There were many channels near these fields, each going to one of the "farm" terraces, apparently serving as carriers for runoff water.

We surveyed over a hundred of these farm systems before we understood how they worked. Each farm unit had two parts: the farm proper consisting of terraced fields for growing crops and trees surrounded by a wall, and a runoff catchment area, from which flood water was carried in channels to the fields. Our survey revealed that the catchment basin was up to 30 times the size of the area under cultivation. As an example, let us suppose that a field of one hectare has a catchment area covering 25 hectares. Furthermore, flood measurements which we carried out over many years showed that the mean annual runoff amounts to between 20% to 30% of the annual rainfall. That means that of 100 mm rainfall, 20 to 30 mm runoff; and therefore if the catchment area of a farm is 25 times the cultivated area, the fields get from 500 to 750 mm of runoff water plus 100 mm in direct rainfall, enough water to support any type of agriculture. This is a theoretical calculation. The farms we reconstructed received in reality, according to the season's rainfall, from 300 mm in bad years, to between 600 and 800 mm in good rainy years. Paradoxically, after heavy floods, we had to let most of the water run out in order to prevent terrace walls from collapsing. Farms without a catchment were of no value, so that the borders of the catchments of each farm were delineated by walls. The water rights of catchment areas were also guaranteed by law, as documents found in Nitzana demonstrated.

We were surprised to find that the terrace walls were only 30 cm above the soil level, retaining only 30 cm of runoff. We were naive enough to believe that the ancient inhabitants were stupid not to have built the wall higher, keeping more water back. Later, the reason for this emerged from our experiments and observations: the depth of the loess soil in the terraces is about three meters, and its holding (field) capacity amounts to 30 cm of water. Had the walls been higher, they would have been in greater danger of collapsing under the impact of large floods. The ancient inhabitants, with their 30 cm walls, had found the correct, equilibrated solution.

They had also invented another anti-erosion method: step-like insets ("drop structures") in the terrace walls, which allowed the surplus water to flow onto the next terrace, progressively reducing the mechanical force of the water flow. The broad spillways in the large and lowermost walls constituted another erosion strategy, enabling the collected overflow of all the terraces to stream out of the farm.

The mechanism of another water-harvesting system presented us with another riddle of the ancient farmers which we called the "diversion system." On one of our flights near the remains of the ancient city of Kurnub we came across a terraced area covering about 112 hectares and therefore much larger than the fields of the "farm units." The total length of its terrace walls was about 250 m with drop structures extending over 30-50 m. We needed many days of surveying

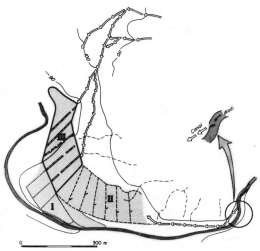

a. Aerial photograph of the diversion system of Kurnub (Photo L. Evenari). b. Field survey of the Kurnub system. *Dotted lines* Remains of walls of Stage I. *Dashed lines* Terrace walls of stage II. *Solid lines* Terrace walls of stage III. *Arrows* indicate the diversion channel leading water from Wadi Kurnub (*heavy black line*) to the system. *Inset* Enlargement of diversion dam in Wadi Kurnub

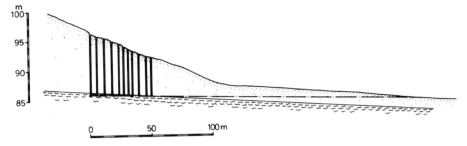

Schematic longitudinal section through a chain of wells. Eleven shafts penetrate gravel and soil to an underground tunnel (*dash-dot line*). Water table indicated by *a light solid line*

until we understood how the system functioned. The city of Kurnub is on a hilltop overlooking one of the largest and longest wadis in the Negev Highlands with a steep canyon. The large watershed area of about 27 square kilometers gives rise to torrential floods in the wadi after heavy rains. These floods could not be utilized directly, but the ancient farmers found a clever way to exploit at least part of the water for their agriculture. At a distance of about two kilometers from the city of Kurnub the canyon opens into a large flood plain, into which, due to erosion, it has cut itself. Exactly at the point where the wadi enters the flood plain, the ancient farmers built a dam diverting a manageable amount of flood water into a channel. In contrast to the smaller channels of the farm units, this channel, leading water into the upper field terraces was solidly built of stones. It was up to 7 meters wide, 1.5 meters deep and 400 meters long.

It turned out that this system was superimposed on an earlier one. We observed faint lines, which seemed like the remains of very old terrace walls stretching from the more recent ones into the flood plain. The predecessors of the people who had built the diversion system already used the flood plain for agriculture. In their time, Wadi Kurnub had not yet cut itself into the formation of the flood plain, and its water flowed freely over it. The peasants of this phase, which we called Phase One, built terraces on the flood plain, the traces of which we had observed. However, at a later stage, after the deposition of a great deal of sediment in the fields by the flood water and the erosion of the wadi into the plain, this arrangement became unworkable. This was the end of Phase One. The diversion system was subsequently developed to deal with the new conditions. This was Phase Two. In time, however, the second system also fell victim to the erosion-deposition cycle. The combination of wadi erosion and sediment deposition on the terraces prevented the flood water of the wadi from reaching the fields. The ancient farmers, by being overly ambitious, had tried to master too large amounts of water without taking topography and hydrology into account. They now developed Phase Three. The lowest lying three hectares were turned into a runoff farm, which no longer received water from the diversion channel, but from smaller channels built on the hillside nearby. This farm remains almost intact to this day. Interestingly similar mistakes are still made by farmers. Indeed, in one of our reconstructed farms at Avdat, we once committed an error of this nature.

Whereas the hydrological-topographical conditions in the Negev

Highlands are ideal for runoff farming, in the Aravah the mean annual rainfall (25 mm) is too low to allow this type of agriculture. The farmers there used an irrigation system based on *kanats,* a chain of walls, also known as *foggaras* or *karez,* which we had already observed by flying over the Aravah. These *kanats* were used to lead groundwater, about 20 meters below the soil surface, which flows from the Negev Highlands to the Aravah, about 1000 meters lower down, and far below sea level. From the base of the hills, the terrain slopes down to the middle of the Aravah. Since at that time pumps had not yet been invented, the ancient farmers used the *kanats* to lead the ground water to irrigate their fields. Near Kibbutz Yotvatah with the enthusiastic help of kibbutz members, we excavated an entire *kanat* system to understand how it operated. It was a difficult task. When we had found the mother well, and its entrance to the underground channel, we danced Israeli folk dances all the way home!

The ancient farmers dug one or several mother wells which extended to the water table at the base of the hills. Down the slope, they then excavated several vertical shafts connected by an underground channel, to lead the groundwater tapped by the mother well down the slope. The shafts were narrow, barely wide enough for one to climb through them in order to keep them and the channels clean. In Persia where *kanats* originated, and some of the ancient systems are still in use, children perform this function. When viewed from the air the excavated soil deposited around the shaft openings looked like bomb craters.

Since the terrain is a downward slope, at a certain point the underground channel eventually reaches the surface, and then continues its course in an above ground channel. On following the latter channel, we found that it led to the chessboard-like areas that we had seen earlier from the air. The chessboard contained rows of regularly-spaced holes. When we visited this area, we found that in each of these holes grew an enormous patch of tussock grass. Its leaves grew out of a stem hard as a stone. The only possible solution to this enigma, seems rather far-fetched. Could each of those holes once have carried a date palm? Did tussock grass, which still grows in Egypt along the bands of the irrigation channels, replace the palm trees after the *kanat* system ceased functioning? The tussock grass must have been several centuries old.

While excavating the *kanats,* we found several ancient Persian and Roman potsherds. This led us to propose, when Palestine was under Persian rule, either the Persians or Jews returning from captivity in Babylon in the sixth century BCE had brought the Persian invention to Palestine. The *kanats* functioned at least until Roman times.

In this chapter, the term "ancient farmers" has been frequently used. We could not complete our survey without knowing who these "ancient farmers" were. Helped by historians and archaeologists we found the answer.

From the start of our studies, it seemed clear that the Nabateans were one of the peoples who practiced runoff farming. Any doubts in this regard were dissipated with the excavation and partial reconstruction of Avdat by Avraham Negev. He found a large stone trough, a libation altar, beside a large diversion system near Avdat, with a Nabatean inscription starting with the words "Dena sichra." He did not know what the word "sichra" meant. Later in a lecture I was invited to give on our Negev work to the Hebrew University's department of Archaeology, I mentioned the dams we had found there referring to them by their Hebrew word "secher." After the lecture, Negev embraced me and said, "You have solved the riddle! Sichra means dams. My inscription reads, "This is the dam, which Garmo and his friends built in the 18th year of our Lord Rabbel, who brought life and delivery to his people." King Rabbel II reigned between 88 and 89 AD. The Nabateans however, were not the first people to settle in the Negev and to practice runoff agriculture. On a trip Liesel and I took alone to Yotvata, we saw a cistern surrounded by a mound of soil near the settlement of Mitzpeh Ramon. The mound consisted of sediment brought in by runoff water that had been stored in the cistern. It had to be removed from time to time by the users of the cisterns to clean them.

The sedentary Negev peoples used such cisterns for storing drinking water, obtained through runoff channels. Interestingly, the Bedouins still use old cisterns to collect water from runoff channels for their sheep and goats in this way. Unlike the Nabatean cistern hewn into the rock, the one we found was open and dug into the soil. I told Liesel that I had an instinct feeling that this was an Israelite cistern. Liesel thought that this was wishful thinking on my part. Closer examination of the cistern revealed that it was lined with large stones to strengthen it. Two channels dug into the nearby hillsides had led runoff into the cistern, which still contained some water. I thought that ruins of houses and potsherds that we found close by belonged to the Israelite period - a view which was confirmed when I showed them to the late Johanan Aharoni in Jerusalem. In addition, when I described my findings to Abraham Melamed, Professor of Biblical History at the Hebrew University, he told me to go and read what was written in the second book of Chronicles 26:10, relating to King Ussiah: "also he built towers in the desert and dug many wells......for he loved husbandry." We were overwhelmed to have come across traces left by our ancestors from the times of the kings of Judah, dating back about 2700 years.

We decided to look for more traces from that period. On our next excursion, we found 26 cisterns of the same type from the same historical period. On subsequent trips, we discovered an Israelite house with an attached runoff farm and an Israelite village surrounded by runoff fields. With Aharoni's help, we excavated both these sites. Later, archaeologists discovered many other traces, such as fields, cisterns and fortresses, from the same period. These findings leave no doubt that the Negev was settled by our ancestors, who practiced runoff farming, from about 1000 BCE to 600 BCE.

A fairly clear picture of the history of the Negev has emerged from research of historians and archaeologists. It was occupied by food gatherers and hunters during the Middle Paleaolithic (from about 60,000 to 30,000 BCE), Upper Palaeolithic (from about 30,000 to 15,000 BCE), Epipaleolithic (from about 15,000 BCE to 8000 BCE), and the Neolithic (from about 8000 to 5000 BCE) periods. During those times the climate was different from today. We found many flint implements including burins, scrapers, knives, arrowheads and hand axes made during these periods in the Negev Highlands. I have hundreds of these including 500 beautiful hand axes, which I collected near Avdat.

No traces of human habitation have been found in the Negev Highlands for the period between 4000 and 2100 BCE. In the Middle Bronze period (about 2100 to 1900 BCE), however, the Negev was settled by people of whose origins and reasons for disappearance remain unknown although they left a great number of turmuli (burial mounds) on mountains. The archaeologist Kochavi excavated some of their villages and came across some agricultural implements, but we did not find any farms or fields from this period.

Apparently the Negev was not inhabited from the end of the Middle Bronze period until approximately 1000 BCE. From then onward, under Solomon and the kings of Judea, the Negev was densely settled by the Israelites. However, this flourishing period of runoff agriculture came to an end around 600 BCE. There was another historical hiatus in the Negev Highlands from about 600 to 300 BCE.

Around 300 BCE, the Nabateans appeared on the historical scene. The Greek historian, Diodorus Siculus (first century BCE), who relied on much earlier reports of Hieronymus of Cardia, a contemporary of Alexander the Great, described the Nabateans as a semi-nomadic people who stored drinking water in cisterns (34):

> "....they lead life of brigandage, and overrunning a large part of the neighboring territory they pillage it, being difficult to overcome in war. For in the waterless region.... they have dug wells at convenient intervals and have kept the knowledge of them hidden from the peoples of other nations. For since they themselves know about the places of hidden water and open them up, they have for their use drinking water in abundance...As the earth in some places is clayey and in others is of soft stone, they make great excavations in it, the mouths of which they make very small....After filling these reserves with rain water, they close the openings....and they leave signs that are known to themselves but unrecognizable by others....While there are many Arabian tribes who use the desert as pasture, the Nabateans far surpass the others in wealth....for not a few of them are accustomed to bring down to the sea frankincense and myrrh and most valuable kinds of spices from what is called Arabia Eudaemon [Arabia Felix]"

In this way the semi-nomads became masters of the caravan trade, carrying expensive merchandise from the East, India and even China to Damascus, Athens and Rome. They built fortresses and cities such as Petra, and in the Negev Highlands, Avdat, Shivta, Nitzana, Khalutza, Kurnub and Rukheiba, to protect their trade routes. From there roads went to the port of Rhino Colura (today El Arish) and Gaza. Apparently, they started runoff agriculture around these cities in order to provide food for their caravans on these extensive roads. They were ruled by sheik-like kings and queens. Most ancient authors, who wrote about the Nabateans, referred to the remarkable fact that they kept no slaves.

In 106 AD, during the reign of their last king Rabbel II, the Romans occupied the Nabatean empire, which, at its zenith went as far as Saudia, Jordan, the entire Negev and parts of Sinai. Nabatea became a Roman province, and when the Roman Empire was divided, became part of the Byzantine Empire. The Byzantines developed the most sophisticated level of runoff agriculture. The Nabatean people remained in the region as the Nitzana papyri indicated (35), but converted to Christianity, and the ancient Nabatean cities continued to flourish.

By 641 BCE, the Arabs conquered the Negev. Although the cities were not destroyed, the population gradually left. As Avraham Negev, who excavated Avdat and Kurnub, put it: "It is clear from the excavation of Byzantine houses that one family after the other packed their belongings and emigrated."

From the eighth century AD onward, the Negev became a pasture ground for the Bedouins. Palmer (33) describes the decay of the old civilization most dramatically:

"Wherever he [the Bedouin] goes, he brings with him ruin, violence and neglect. To call him "son of the desert" is a misnomer; half of the desert owes its existence to him, and many fertile plains from which he has driven its useful and industrious inhabitants became in his hands like the South Country [the Negev], a parched and barren wilderness. Where scores of thousands once lived in prosperity a few hundred Bedouins barely eked out their subsistence. Farm systems were left untended and allowed to fall into a state of decay. The life-giving floods now became a destructive force, gradually eroding away the terraces. Thus, the tremendous and sustained effort of many generations was wiped out through human neglect and abuse. "Long ago, the word of God had declared that the land of the Canaanites and Ammorites should become a desolate waste; that the cities of the South shall be shut up and none shall open them" (Jer. 13:19), and here around we saw the literal fulfillment of the dreadful curse. Walls of solid masonry, fields and gardens encompassed by goodly wells, every sign of human industry was there, but only the empty names and stone skeletons of civilization remained to tell us what the country once had been. There stood the ancient towns, still called by their ancient names, but not a living thing was to be seen, save when a lizard glided over the crumbling walls, or a screech owl flitted through the lonely streets."

Chapter 12
An Orchard Grows from Ancient Seeds

August 26, 1956 presented us with a blazing hot dry *khamsin* (hot desert wind) day. That morning, we surveyed a farm near Shivta, and at around midday, when the heat reached 108 degrees, we decided to stop work. We put blankets over the jeeps to give us some shade while we ate lunch, Liesel's famous desert salad. While discussing our ideas about how the farm might have worked, completely out of the blue, Liesel said, "Why don't you stop theorizing? You are experimental scientists, why don't you test your theories by reconstructing an ancient farm?" As if struck by lightening, none of us had ever thought of that. Our weariness disappeared and we began to think about finding a farm to reconstruct. Looking at the farm we had just surveyed, Kofish declared that it was eminently suitable for such a purpose. He was right. The farm was well preserved with terrace walls and channels, and not too large, with a clearly delineated catchment area. The ruins of a large farmhouse on its upper reaches with an intact cistern and a runoff channel still containing water, particularly attracted Liesel, who planned that we should live in or near the farm and was fascinated by the idea of rebuilding the farmhouse in which she wanted us to live.

On our return to Beer Sheva, she found an architect who subsequently came to the farm, measured the house and suggested how we could go about reconstructing it in the original style. Leslie prepared a detailed plan for reconstructing the farm while I tried to find the necessary financial means for that purpose and the scientific work we wanted to carry out there. We were fortunate to be awarded grants by the Ford, Rockefeller and Rothschild Foundation. A year later we started our work at the Shivta Farm.

At the same time, an Israeli government team lead by Jossi, who was a member of the nearby kibbutz Revivim reconstructed the town of Shivta. His workmen were new immigrants. Jossi was enthusiastic about our intentions and offered to carry out Leslie's plan. Initially, all went well. He was, however a very obstinate man, thinking he knew everything better than other people, and we eventually came to a head-on collision with him. After careful calculations, Leslie told Jossi how high the critical lower wall and how broad its spillway should be so that the overflow during strong floods would flow out of the farm without causing any damage. But Jossi knew better, he thought that building the walls 50 cm higher would make more water available for the crops. Defying Leslie's orders he carried out his own plan. Refusing point blank to make the wall lower, a stormy conflict developed between the two men. Jossi declared that he would forcibly evict Leslie from the farm if he dared to return. The first flood we experienced, which completely destroyed the wall, proved how right Leslie was. At that point, we dismissed Jossi. After reconstructing the farm we wanted to go ahead and rebuild the farmhouse. However, Avraham Joffe, a friend of mine with whom I had fought together during the Second World War, who was in charge of the army units in the Negev at the time, put us off, telling us that it was dangerous to live in Shivta because he could not guarantee our safety.

Shivta, four kilometers off the main road from Palestine to Egypt, which had fallen into disuse after the War of Independence in 1948, is in a very isolated position. There were many marauders from the neighboring Arab countries in the area, one of whom had recently murdered a young army officer near there. So, we gave up our plans to live in Shivta, and only used the reconstructed farm for flood and climatic measurements and agricultural purposes.

Liesel was convinced that proper scientific work on a reconstructed farm would only be possible if some of us lived there to observe the unpredictable rainfalls and floods. Therefore we looked for another farm in a relatively safe area. By chance, we soon came across a suitable place.

In addition to the project at Shivta, another government team under Avraham Negev was reconstructing the city of Avdat. Jehoshua Cohen, a member of Kibbutz Sde Boker, who was in charge of the work team there, took us to the top of a hill opposite to view the work in progress. We told him about our frustrated plans to live in Shivta. "Look down the hill," he said. "There are terraces of an old, relatively well-preserved farm. Why don't you reconstruct it? You could live safely on the farm since the road to Eilat passes by the farm, and the new town of Mitzpeh Ramon is not far away. The army would surely be able to guarantee your security here." Although there was no farm house on the site, the idea seemed to make sense. We surveyed the farm, found it suitable for our purposes, and decided to go ahead with the reconstruction.

We started work a few weeks later in the summer of 1959. With permission of Teddy Kollek, who was Director of Ben Gurion's office, who was Prime Minister at that time, we employed as workers new immigrants from Morocco, Tunisia, India and Persia, who lived in Yeroukham, a village about 30 kilometers north of Avdat. We were fortunate to find Eddie Dribben, a friend of Kofish and put him in charge of the rebuilding.

A real non-conformist, whom we nicknamed "the Jewish Cowboy," Eddie Dribben's father was the sherif of a small town in the West. His wife, who had been a partisan in Russia, came to live in Israel in 1945, and was the commander of an artillery unit during the War of Independence; she was wounded and both legs were paralyzed.

Eddie had "golden hands" and was an expert stonemason. He never used prefabricated building utensils, he forged all the tools he needed in his own smithy. In the course of the work he trained several completely unskilled immigrants to be good stone masons, giving them a useful trade, and one of them, Khananya, worked with us for many years. Unlike Jossi, Eddie followed Leslie's instructions to the letter, and the terraces and walls he built are still standing now, 29 years later.

I now made the fateful decision to devote most of my time to my research on the farms. I resigned as Vice-President of the University, feeling that with 32 buildings on the new campus in Givat-Ram, I had done my duty. I also realized that if I were to stay on in that tough administrative position, I could no longer continue with my career as a scientist. During the reconstruction of the Avdat Farm, I only spent three days a week in Jerusalem in order to give lectures, the other four days, Liesel and I stayed in Yeroukham in a small, two-room house built for new immigrants. From there, we drove down to Avdat. Shimon, a young Jew from Yemen stayed there permanently in a trailer to guard the farm.

We soon learnt how unpredictable the desert can be. One day, while in Yeroukham Liesel was preparing the usual festive meal for the Shabbath Eve, it started raining and since we were worried that this would mean a flood in Avdat, taking our half-cooked meal with us, we rushed over to the farm. However, there it was a cloudless, star-covered night and not a drop of rain fell over the next two days. On November 10, 1959, when we least expected it, a terrific downpour suddenly came forth from a cloudless blue sky - the first torrential flood we had ever experienced. The farm was completely water logged.

Although the reconstruction work on the farm was not yet complete, we decided to use the flood water to grow barley. We did not yet have any farming implements and therefore asked the sheikh of a local Bedouin tribe to send over two or three men to plough and sow the fields. To our amazement, 21 men came and not wanting to offend them we let all of them work. With a great deal of singing and shouting, they went about their task. One plowed from right to left, his neighbor from left to right. Two weeks later, the first green sprouts broke through the soil - our real work had begun. In the spring of 1960, we reaped our harvest of 1.25 tons of barley per hectare which was a good yield considering the primitive plowing and sowing methods we employed.

Although reconstruction of the farm was now complete, we still had no house in which to live on the site. Clearly we could not continue to commute daily from Yeroukham to Avdat. In spite of having no money for building the house, we asked Zoltan Harmath, a Jerusalem architect, to draw up a plan. Neither the Hebrew University nor any of the Foundations who had helped us to reconstruct the farm were prepared to finance the building of a house. As it turned out, another chance occurrence solved our problem.

In the Spring of 1960, our friend Sam Risk, the Director of the Office of the Canadian Friends of the Hebrew University, sent Miss Ann Lautermann from Montreal to visit our farm. A petite, delicate, shabbily dressed, eighty-year-old woman, using a walking stick, she arrived on a hot, *khamsin* day. Since we had no trees to provide shade, we had lunch in Nasser's tent, a Bedouin with whose children Miss Lautermann struck up an immediate friendship. We talked to her about our work also mentioning our "housing problem." All of a sudden she said: "You cannot go on without a house. Just let me know the cost involved and I will cover it." We could not believe our ears. From where would Miss Lautermann obtain the money?

Risk later told us that Miss Lautermann was a multimillionaires. The daughter of a rabbi from Eastern Europe, she had come to Canada with her family. A school-teacher by profession, she inherited a small sum of money from a brother, and by playing the stock market, she made a lot of money. She spent very little on herself, making all her own clothes, and became a great philanthropist. Subsequently she donated a large building on the Givat Ram Campus in honor of her parents.

After we got the money, Liesel set to work on the house. She supervised the building according to Harmath's plans and got permission to use the beautiful ancient hewn stones which had fallen from the Nabatean city of Avdat. Consequently, the house was built entirely of stone. It was designed primarily as a functional building with not a wasted square meter. Liesel, helped by members of our team, made all the furniture in the Swiss peasant style. It is so solidly built that we still use it today. On November 29 we moved into the house.

We were sorry to leave Yeroukham, and to have to break off contact with the new immigrants, mostly originating from Arab countries. The older generation had found it very hard to get accustomed to a new country with a completely alien culture. They did not speak Hebrew. Some of them were so primitive that they believed the W.C. was meant for washing dishes. We helped them as much as we could. In contrast, it was interesting to see how fast their children adapted to their new society. The government did very much for the newcomers, providing every family with a new house which had a small garden with a few fig trees and vines and ready made food, mostly canned goods. Liesel often had to teach them what to do with them. This and bread tide them over the first few days. The men were immediately given jobs and that is how we got our work force for the reconstruction project. Some of them became members of our permanent staff, which we needed when the farm was more or less completed.

Most of the staff consisted of young Israelis who were prepared to live in the Negev. When we were still in Yeroukham, they lived in an immigrant house opposite ours. When the house in Avdat was ready, they moved in with us. We were a strange team. It takes a certain nonconformist nature to be prepared to go and live in the middle of the desert far from the comforts of city life. Shimon, the young, wild Yemenite guard who had lived in our van in Avdat was one of them. Every week, when Liesel distributed clean sheets, he refused to use them and threw them under his cupboard. Liesel tried explaining that it was easier to wash sheets than blankets, but he was deeply hurt. "I", he said, "am a clean man and do not dirty my blankets. From now on I am not going to sleep in the house any more." From then on he slept on the flat roof of our house. His physical abilities were quite remarkable. From a standing position he could jump directly onto the roof of our house. None of us were ever able to repeat his performance.

Our team consisted of five people, including two *Sabras* (native-born Israelis). As long as they had lived in their own house in Yeroukham they hadn't disturbed us. Now, living together in a relatively small house, we could not get accustomed to their disorderly habits. In an attempt to improve the situation, we took on a young kibbutznik to take charge of the work team. Liesel explained his duties, asked him also to take responsibility for maintaining a modicum of order in the house. He turned out to be a fanatical socialist. In one of his first crusades, he tried to persuade the Bedouin watchman Jumma to stop praying five times a day to Allah according to the Islamic custom, since there is no God. He did not succeed, but was successful in convincing him not to give part of his monthly salary to the sheikh of his tribe. This is an ancient Bedouin custom, a kind of recompense for the sheikh who helps his people whenever they are in conflict with the government. Jumma really stopped making payments to the sheikh, who replaced him by someone else who was willing to keep to the rule. When we reproached the manager, he explained that he was an atheist and a socialist; it was his duty to teach the Bedouin a new social order. I tried to point out that his actions might endanger our good relations with the Bedouin.

Our "Savonarola" as we soon called him then decided to go further and apply his new social order to our team, declaring that all our workers were free men and were not obliged to keep to the orders of the house. They could do whatever they wanted in their own rooms. He refused to see reason, and said that Liesel behaved like a Prussian officer. To make things worse, he introduced a new working scheme. They now worked from 6 a.m. until midday, lunched and then slept until 6 in the evening. They then made a bonfire, which they sat around until 2 a.m. Consequently, our sleep was disturbed. The matter came to a head when the team refused to come and help us to unload our jeep when we arrived from Jerusalem. One day arriving at Avdat, we entered the house and found them eating in the kitchen. We asked for help. Nobody moved. "Savonarola" had passed the word around that we were exploiting the workers and they were not obliged to help us. I was nearing a breakdown, and seriously considering giving up the project altogether. Once again Liesel stepped in with a radical solution: she dismissed the entire team and replaced it with a better group with which we had no more trouble!

Initially our primary aim in the reconstructed farms was to test our runoff theory, which we did by installing automatic flood gauges where channels came into the farm. It soon became clear that for understanding runoff formation we would need climatic data, particularly relating to the rainfall. So we put up meteorological huts and many rainfall gauges of various kinds in Shivta and Avdat. As a result, we have data from 1960 to the present, presenting an uninterrupted record of the climate in the area and we have been awarded many prizes by the Israel Meteorological Service for providing them with information.

Aerial view of the reconstructed Avdat farm. In the *foreground* are four unreconstructed terraces. The channels leading runoff from the hills are clearly visible. On the hill to the *right* is the farmhouse (Photo L. Evenari)

Flooded fields on the Avdat farm (Photo L. Evenari)

The orchard in Avdat. On the hill are the ruins of the town of Avdat
(Photo L. Evenari)

Agro-forestry plot in Wadi Mashash. Sheep grazing on grass grown between trees (Photo Z. Loewenstein)

The first Israeli cistern we found (Photo L. Evenari)

After our success with the first barley harvest in 1960, we became more ambitious and decided to find out whether the ancient agriculture methods could be of practical value today. We became "farmers" although this had not been our original intention. In addition, as a botanist and plant physiologist, in cooperation with several of my assistants and my former student and friend Yitzhak Gutterman from the Hebrew University, we gradually turned Avdat into a center for research on the ecology of desert plants.

We learnt a great deal from our meteorological measurements, the rainfall data. Most important, we found out that unpredictability of the rainfall, the "uncertainty syndrome," as we termed it, had quantitative, temporal and topographic dimensions. Thus, in the winter of 1962/63, the driest year we experienced, only 25.6 mm of rain fell, producing only a small flood in spite of which our crops survived. The next year we recorded 183.3 mm, the largest rainfall we ever had in Avdat, and the farm was flooded many times. The total duration of the rainy season and timing of the first and last rainfall also do not show a fixed pattern. In 1960/61, rain only fell for four months, while in 1963/64 the rainy season lasted seven months. In 1965/66, the first rains fell on October 5, and in 1962/63, on January 17 - a difference of 103 days. In 1960/61, the last rains fell during February, and in 1966/67 on May 10, 1967 a difference of 85 days. The rain was also highly localized, so that the main road near Avdat was sometimes half-dry and half-wet. Desert farmers have to contend with all these uncertainties and act accordingly.

We limited our agricultural activity to growing fruit trees, including apricots, apples, almonds, pistachios, olives, vines and cherries, field crops and pasture plants. Whereas apples and cherries didn't do well, apricots, peaches and vine gave relatively high yields and the fruit was sweeter and had a better flavour than those grown under irrigation in the north of the country. Lady Rothschild, on a visit to Avdat, ate some of our peaches, and said that no other peach would ever taste good to her. But we could not compete with the irrigated fruits from other areas because their yields were much higher. However, almonds and pistachios, both yielding about 1.4 tons per hectare, and olives, at up to 10 tons per hectare proved to be well suited to runoff desert agriculture. These yields are similar to those in other parts of Israel and various other countries. Pistachios proved to be best adapted to desert conditions of all the crops we tried.

We also grew wheat, barley, peas, onions (for seed production), artichokes, asparagus, sunflower, safflowers and chick peas, with an average per hectare yields of 2.7 tons for barley, 3.6 for wheat, 1.9 for sunflowers, 2.5 for safflowers, 5.5 for peas. These yields are high for desert areas especially of developing countries but cannot compete with the higher yields obtained with irrigation. In a long-term selection experiment, we tested 46 different species and varieties (cultivars) of annuals, and 81 of perennial pasture plants. Of the annuals wild oats, medics and vetch and of the perennials alfalfa, smilo and harding grass proved to be best adapted to desert conditions. These agricultural results show clearly that runoff agriculture can contribute significantly to the alleviation of hunger in desert areas of developing countries.

In this and the previous chapter no explanation was given as to how runoff in deserts is formed. After a certain amount of rain has fallen, the upper crust of loess, typical of the Negev Highlands, forms a very thin crust which is nearly impermeable to water. After the crust has been formed and the rain continues, most of it does not penetrate the soil, but runs off the hills into the plains and wadis. Over several years, we measured the hydrological effect of this crust. Prior to its formation, about 18 mm per hour of water infiltrates the loess. However in the presence of the loess crust the infiltration decreases to an hourly rate of only about 3 mm penetrating the soil. Any rain which produces more than these 2-3 mm per hour runs off, causing floods. The mean annual runoff is up to 30 % of the total annual rainfall. Another feature of considerable practical importance emerged from our studies. The eight channels for carrying runoff into the terraces at the Avdat farm, each have catchment areas of different sizes. With the use of automatic flood gauges at their point of entry into the farm, contrary to our expectations, we found that the smaller the catchment areas, the larger the runoff per one square meter. Thus, a flood on December 10, 1963 produced 1238 cubic meters of runoff water from a catchment area of 340 hectares, meaning it produced 1238 : 340 = 3.6 cubic meters of water per hectare; whereas corresponding figures for a three-hectare catchment area were only 156 cubic meters of runoff. In this case each hectare produced 156 : 3 = 52 cubic meters of water. The underlying reason for this phenomena is known as "overland flow" in hydrological terminology. To put it simply, the further the runoff has to travel to the farm, the more water is lost due to depressions, stones and other irregularities of the surface of the catchment. For this reason sometimes after moderate rainfall the small catchments gave rise to floods, whereas the larger ones produced none.

This gave us an idea. Why not grow each tree or crop in its own small catchment area, which we called "microcatchment"? We divided the more or less level flood plain opposite the Avdat farm into small units ranging between 1000 and 16.5 square meters. We did this because we did not know what the optimal size of microcatchments would be for various crops. We then put up earthen borders, about 20 cm high, around each plot. A tree or a shrub was then planted in a small basin at the lowest point of each microcatchment. Thus, the basin received only the runoff from its own plot. The experiment was a success. We found that under the prevailing rainfall conditions, the optimal microcatchment size in the Avdat Farm for fruit trees was about 200 to 300 square meters, for vine 150 to 200 square meters, and for salt bush (a fodder crop for cattle), 100 to 125 square meters. Clearly, in other loess-soil arid areas with greater rainfall, their size would be much smaller.

Although we were not aware of it at the time, we were not the originators of this idea; we subsequently found that peasants in Southern Tunisia grow olives in microcatchments, a technique that might have been introduced to the region by the Phoenicians who built Carthage. The advantages of this method are obvious - crops get more water this way than their counterparts on the farm. In addition rains which are too small to give rise to floods on the farm do so in such microcatchments. It is also much cheaper to build and maintain microcatchments in good repair than large farms, and they can be made in various shapes and sizes to suit the topography. Last but not least it seemed that such a simple technique could easily be introduced into developing countries.

The Avdat farm was an experimental enterprise with only a few hundred trees. However, proving the value of runoff farming unequivocally, required further extensive tests on a much larger scale. For example, how could we be certain that our best selected pasture plants were really good cattle feed before we had tested them in a grazing experiment with sheep?

Once again, an "accident" intervened to solve our problem. After the Six-Day War of 1967, we had again free access to the Hebrew University on Mount Scopus. The Botanical Garden there, including a small forest of Lebanon cedars, had been partly destroyed, and many trenches had been dug there by our army. A group of German volunteers from the Sühnezeichen (Action Atonement) organization, which has as its aim a desire to atone for the sins of the Nazis, had come out to help repair the damage on Mount Scopus. I told Otto Schenk, the leader of the group, and his friend and co-worker Udo Nessler about the need to construct another larger runoff farm. They were enthusiastic and offered to send some volunteers to help us with the work, and to try and raise the necessary funds. The money for the venture came from four sources: Prof. Freudenberg, "Heks" (The Evangelical Aid Society of Swiss Churches, whose director was Hans Schaffert), the "Bread for the World" Organization of Germany, and the Evangelical Church of Hessen Nassau, whose President was Helmut Hild.

Although we were ready to start, we had not yet located a suitable site in the Negev Highlands, large enough for our purposes. We again took to our jeeps and after several weeks of driving around we decided on Wadi Mashash, a broad valley, about two kilometers wide, about 20 kilometers south of Beer Sheva. The farm and its catchment extended over an area of 20 km^2.

However, we then ran into difficulties. Part of the site was used as grazing land for sheep and goats by the Bedouin Asasme tribe; and another section was an army training area. The main difficulty concerned the ownership of the land. The Asasme, under their sheikh Odeh Abu Muamar did not own land there, which legally belonged to the State of Israel, but only had grazing rights. After seemingly endless negotiations, we eventually came to an agreement: the army would give up the area in favour of the Hebrew University, the land was leased to the Hebrew University for a three-year period, with an option of renewal, and we came to a private arrangement with the sheikh, allowing his tribe grazing rights in certain parts of the Wadi.

During our negotiations with the Bedouin, we often visited in Odeh's tent, and he also came to our house. The large tent in which he entertained his guests was made out of black sheep's wool. Only non-Bedouin women were allowed to sit with the men. His three wives each had their own tents, which were out of bounds to men. Liesel often visited the women and told me that the youngest, a 17 year old, was a real beauty.

It always hurt Liesel's feelings, who is against killing animals, to see how a young goat was slaughtered and the meat cooked when we arrived as guests. During her visit to the women they prepared pittah (flat bread), which is similar to Matzot (unleavened bread), that Jews eat on Passover. The dough made out of flour was poured onto a hot metal plate, and was ready within a few minutes. The pile of pittah was then thrown on the ground and some of them became covered with what looked like pieces of charcoal but were animal droppings. Liesel got back to the men's tent and warned me in a whisper not to eat pittah from the bottom of this pile!

In the main tent, sweet tea in finjans (small cups) and then coffee, freshly ground in a small wooden mortar and pestle, which made pleasant musical sounds as it was moved in a regular rhythm and flavored with "hell" (cardamon), was served. I can still hear the regular rhythm which never changed. Then the meat was brought in a large dish. The sheikh selected the best pieces with his hands and offered them to me, his guest of honor. Since we were not accustomed to eating fresh unsalted meat, it gave us mild indigestion. On future occasions we carried salt in our pockets and used it surreptitiously so as not to offend our host. These visits always took hours. We would first discuss the weather, and various farming matters, and only then the business on hand. After the meal, we all washed hands, and then the sheikh, or one of the older members of the tribes would sing old Bedouin songs or improvised ones in praise of the honored guests, to the accompaniment of a rubaba (a string instrument). In appreciation for being entertained we always brought a lot of sugar and tea according to the Bedouin custom, in addition to some clothes for the children and trinkets for the women.

In the summer of 1970, before concluding the negotiations, the volunteers, under Schenk and Nessler, already started construction work in Wadi Mashash. We planted about 3000 olive and almond trees, and a few fig trees and vines, some in microcatchments and some in an experimental structure we called "limanes" (the Greek word for "lake"), which are small tributary wadis with earthen check dams, to retain runoff, resembling small lakes when flooded. They proved to be very successful. We turned a section of the wadi extending over 50 hectares into good pasture by replacing the local vegetation, which the sheep would not touch, by the selected pasture plants we had tested in Avdat. Our flocks consisted of Awassi sheep, the breed used by the Bedouin, and we showed that with controlled grazing, this pasture could support far more animals than those of the Bedouin. We got water for the sheep from an old well which was completely filled with sediment, and which we excavated after an old Bedouin had shown us where it was.

Apart from shortage of food, another major problem in the arid zones of Africa is caused by cutting down trees to provide fuel. Indiscriminate felling of trees leads to soil erosion. We investigated two possible approaches toward dealing with this situation: We planted trees which could be used in Africa as fuel and pasture trees in runoff limanes, and a strategy based on agro-forestry by planting rows of trees between alternating rows of annual crops.

We also enlarged the Avdat project in 1977 by planting thousands of pistachio and almond trees at a site opposite the farm in order to carry out a large selection experiment for drought-resistant and high-yield varieties (cultivars). David Masig, who was in charge of the Avdat farm at that time, suggested that we get water from Wadi Avdat by building a 400-meter long channel from the nearby wadi. When it flows, it carries enormous amounts of flood water. I hesitated to agree to David's scheme because I was afraid that we would make the same mistake the ancients had made in Kurnub. I finally agreed, however, because I wanted to see if we could, by utilizing modern methods, find ways to overcome the danger of erosion. The new farm was named in honor of Kofish, who died of lung cancer in 1973. The trees in the new plantation grew well but we failed exactly as the ancients had. Already, during the first flood, the earthen channel was dangerously eroded. The force of the floods destroyed most of the terrace walls within a short period.

Later that year during a visit to Saarbrücken, I was hospitalized with a serious infarct. Since the doctors were not too certain of prospects of recovery, David suggested building a 1.3 hectare desert runoff park named after my family next to the Kofish farm. In the middle of the proposed park, he had found a large block of stone resembling a lion, a Lionstone.He thought that this was a sign to name the park "Loewenstein" park since this is the English translation of my name.

Although all the settlements, kibbutzim, villages and towns in the Negev have green areas and parks, they are all irrigated by water brought in from outside the area. David wanted to prove that it was possible for a park that utilized only runoff water to thrive in the desert, with shade giving trees and areas that would remain green even in the summer. My family in Detroit, notably my second cousin Phyllis Loewenstein, provided some of the money for this project, and we planted about 400 trees, 5000 bushes and ornamental herbs. Now, after only eleven years, the park is beautiful and some of the trees are over 12 meters in height.

Today, Avdat also serves as a training center in runoff farming for teams from developing countries. We started this activity by giving courses to a group from Kenya and to people from developed countries who were ready to serve in deserts of the Third World. As a result of this program, there are now runoff projects in Kenya, Burkin a Faso, Niger, Mexico, India, Australia, in the reservation of the

Navajo Indians in North America, and in a neighboring Arab country. Some were established with our help and others based on an imitation of our work. The largest project of this kind was initiated in Afghanistan on a 70,000-hectare site by a German team. They reported that their runoff was a great success, but today after the "troubles" in that country I fear that the project was destroyed.

The practical importance for Israel and other countries with similar deserts of our findings relating to runoff farming, are considerable (36). As a report published by the World Bank in 1985 puts it: "Runoff harvesting is potentially one of the most important techniques for improving plant growth and reducing erosion in semi-arid areas."

Chapter 13
The Six-Day War and its Aftermath

On the morning of May 17, 1967, our handyman Makhluf, a new immigrant from Morocco, helped us to pack the jeep. His parting words to us as we drove off to Avdat were very strange: "My rabbi told me today that we will soon be able to walk through the streets in the Old City of Jerusalem." Since the Old City had been in Jordanian hands since 1948, and there seemed little hope of any change, Liesel laughed off his remark: "Neither you nor we, or even not our grandchildren will have that pleasure," she said. But the rabbi was right. When, after our return from Avdat, we saw Makhluf, we had already been to the Old City.

On the way to Avdat we saw many army trucks carrying men, some in uniform and some in civilian clothes. We had not read the newspapers or listened to the radio that day, and had no idea that the reserve soldiers had been mobilized. Only when we got to Avdat, did we realize what was going on. Only one French volunteer was left on the farm, the rest of the team, Michael Sadeh, Aryeh Rogel, and David Masigh had been called up.

The two weeks between the mobilization and the outbreak of war were nerve racking for us in Avdat. Avdat is near the Egyptian frontier and our situation was serious because we knew that the Egyptian forces were concentrated only 30 kilometers from Avdat along the border in Sinai. Drawing on my army background, I knew that they would try to break through to Beer Sheva, going past Mitzpeh Ramon and our farm. Indeed, documents discovered after the war showed that was what they had intended. My son Eli tried to convince us to leave the farm. We refused, remaining there with David Masigh, who was rejected by the army because he had several damaged fingers, and Ziona, our cook, from Dimona who drove down to Avdat every day even though most of the people in Dimona and Yeroukham were down in the shelters.

Once our own army gave us cause for alarm. A seemingly endless line of truckloads of soldiers, cars and tanks drove past the farm in the direction of Mitzpeh Ramon with their lights full on and making a terrific noise. I thought that they must have gone mad, to reveal their position to the Egyptians in this way. By the next morning,however, all the army units had disappeared. After the war we found out that this had been a successful attempt on the army's part to foil the Egyptians.

The war began in earnest on June 5. That day, I remember picking boysenberries (a cross between raspberries and blackberries) to the sound of cannon fire. We were worried about our son Eli and his family and about my sister Erna in Jerusalem, which the Jordanians were shelling. For a long time our phone calls went unanswered, and we only eventually reached her when she emerged briefly from the shelter to see if her flat was still intact. We were relieved to find out that she was safe but worried about Mickey and Aryeh, who were in the midst of the fighting.

Initially, we could not see how a small country with two and a half million people could possibly win a war against the 60 million Arabs. However, the morale of the population was nevertheless kept high by the daily broadcasts of Chaim Herzog, who is now the President of the State of Israel. He talked to us on the radio, explaining frankly the current situation. We always listened to him in our darkened room and felt much better after his broadcast. As the sound of cannon fire faded, it became clear to us that we were winning the war.

On June 20, we returned to Jerusalem and experienced the indescribable feeling of entering the Old City that had been inaccessible to us for 19 years. Jerusalem's mayor, Teddy Kollek, decided to open the gates of the Old City straight away, even though some Jordanian soldiers were still hiding in the old houses there, in the hope that immediate contact between Arab and Jewish inhabitants of the town would facilitate reunification of the city.

When we entered the Old City, a class of children and their teacher were walking in front of us singing Psalm 126: "When the Lord brought back those that returned to Zion, we were like unto them that dream. Then our mouth filled with laughter and our tongue with singing." These eloquent words expressed what we felt. We were deeply moved by the Western Wall, the remains of our ancient temple, which Liesel had never seen. I have always disliked the expression "holy places," because no *place* can be "holy," and the belief in their "holiness" has caused so much war and bloodshed over the years. However, touching these ancient stones vividly brought home to me my links with the long chain between the Jews of the past and our own generation.

I visited all the Arabs in the Old City, whom I knew before 1948. They received me like an old friend, embracing me according to the Arab custom. Both Arabs and Jews felt certain, that there would now be peace between our nations. How wrong we were! For ironically today, over 21 years later, peace seems further away than it did then.

The next day we went to Bethlehem to try to locate Ibrahim Kattan, my former landlord. However he had died in the meantime, but we did renew ties with his family.

Shortly after the war, Kofish came back from his service in the army. He pounced on me with the idea that we had to go and investigate the plant and animal life, and geology of the Sinai, which was not well known at that time. He considered this a top priority, since we did not know how long the Sinai would remain in Israeli hands, and should therefore not let this opportunity slip by. Together with the Hebrew University's Department of Geology and Zoology we immediately started organizing an expedition to the region.

Kofish's enthusiasm was contagious. The next day I went to see Moshe Dayan who was Minister of Defence at that time and whom I knew well. He immediately agreed to our plans and assigned a special military "Refael" unit under the command of El Yashiv, to arrange the logistics of our expedition. He also agreed to give us army command cars, specially adapted to desert conditions, and to cover the expenses of our expeditions. We did not hesitate, by July 7, we were in the Sinai. The only civilian car was our jeep, which Liesel drove. Our group consisted of Leslie, Kofish, Gideon Orshan and Avinoam Danin of the University's Department of Botany, several geologists under Jaakov Ben Tor, and various people from the "Refael" unit. The entire Sinai Desert lay before us and we could roam anywhere to our heart's content.

Since it was only three weeks after the War, it was still somewhat dangerous since there were many mine fields, Egyptian soldiers were still hiding there, and the area was littered with burnt-out tanks, cars and trucks, anti-aircraft and anti-tank artillery, multi-barreled Katyushas and rockets. El Qantara on the banks of the Suez Canal had been completely deserted by its inhabitants, and we gave water to a lonely, thirsty horse, roaming the streets there.

Once, we slept near an Israeli army unit carrying ammunition. In the middle of the night we jumped out of our sleeping bags. All of the guns around us had started to fire at several Egyptian boats trying to cross the Suez Canal. The Egyptians retaliated by shooting back; we were afraid that some of their shells would hit our ammunition column. Liesel, who had never witnessed an artillery duel before, reproached me: "I can understand that you are prepared to risk your life for Botany, but are you also willing to put me in danger? I think I'll leave tomorrow." By the next morning, however, after the situation had calmed down, Liesel had changed her mind and nothing could have kept Liesel from continuing the expedition; in spite of the danger involved she had fallen in love with the Sinai and with our adventures.

We collected plants and rocks to our hearts' desire across several not previously known areas. A few plant species we found were new to science. We all, especially Liesel, were overwhelmed by the majestic, monumental and unique beauty of the landscape.

Kofish, Leslie and I had another objective in mind. We searched for traces of ancient runoff agriculture, finding many, particularly in the broad El Arish Wadi and its tributaries. Would this not be an ideal place to settle the Palestinian refugees?

We also went to see the Monastery of St. Catherine where the monks were very friendly and hospitable. What a contrast to our visit there in 1940 when they refused to let us in. Shortly after we got there, the monks rang all the bells; at first, we didn't know why. We then saw Moshe Dayan getting out of his helicopter, the monks welcomed him like a king. He was pleased to see us. We talked to him about the agricultural potential of the Sinai. He listened to us attentively. When he greeted Liesel, she said to him: "The first Moshe (Moses) climbed Mount Sinai, and the second one came down on to the mountain." Dayan was amused by the joke.

The monks let us eat the customary Friday evening meal in a large hall. We recited the traditional blessings over wine [Kiddush] and two loaves of bread - probably for both the first and last time within the walls of this Byzantine sanctuary. Joining Dayan, Liesel and I had the occasion to see the well-known unique library, which was only opened for important visitors. We also saw their famous collection of beautiful Byzantine icons, which had been saved from the Byzantine iconoplasts in this isolated setting. They had acted according to the words of the Ten Commandments: "Thou shalt not make unto thee any graven image, nor any likeness of anything that is in heaven above, or that is on earth beneath or that is in the water under the earth." (Exodus 20,3).

The day after our return to Jerusalem, Dayan called to ask me to fly over Sinai with him on July 27. He wanted me to point out the areas suitable for agriculture, which were potential sites for settling the Palestinian refugees from the camps that had fallen into our hands in the Six-Day War.

Liesel, who wanted to photograph during the flight as usual, came with me to the military airport, but Dayan politely told her that she would have to stay behind because he had also forbidden his generals to take their wives on the flight.

In the helicopter, Dayan lay down and immediately fell asleep. After a short flight, we arrived at another military airport, where some other generals joined us. The helicopter was already airborne when Dayan woke up and realized that one of his generals had brought his wife along. He ordered the pilot to return to the airport immediately. "I did not even take Mrs. Evenari, who wanted to take photographs. Tell your wife to get out." Having settled the matter he closed his eyes and went back to sleep. Without further discussion, the pilot turned back and "Mrs General" left us.

Another civilian, a geologist, was also with us. He was supposed to point out the sites with valuable minerals. We both directed the pilot, Mordechai Hod, the Commander General of our Air Force at that time. On the return journey Dayan told the pilot to fly to the old Egyptian temple of Serabit El Khadem. Hod circled over many mountain tips, which all looked similar, but could not find it. Dayan became impatient, and ordered the pilot to land in a Bedouin camp. He got out of the helicopter and asked an old Bedouin if he knew where Serabit was. When he answered in the affirmative, Dayan politely asked him to come with us and show us where to go. The Bedouin sat down next to the pilot, and as if he was accustomed to flying, in no time at all he directed us to Serabit El Khadem.

While we were admiring the remains of the Temple, Dayan led several soldiers carrying a stretcher over to one of the steles bearing a hieroglyphic text, and ordered them to load it on to the stretcher and carry it to the helicopter. At that time, I did not know about Dayan's "passion" for archaeology, and had no idea why the stele had been taken, but to his credit, he later replaced the stele at the original site.

On the return flight, the pilot wanted to land anywhere to get rid of the Bedouin. "No," said Dayan. "You fly back to the camp from where we took him." When the Bedouin left the helicopter Dayan gave him food items as a reward, and we parted from him with many "masalami" (with peace) and "Allah yisalmak" (Allah may give you peace). Did he ever fly in a helicopter again? I am sure that he must still sing about that experience to his grandchildren around the evening camp fire!

For the next three years, members of the botanical-soil science and geological teams spent about two weeks a month in the region. We investigated every wadi, even the most inaccessible ones. We know now every nook and cranny of the peninsula. However, it was not only the scientific work which made Liesel and me "Sinai-crazy". We were captivated by the Northern Sinai plains, the regs covered by black, shining stones, seemingly stretching into infinity over which we could race our jeep at the highest speeds. Surrounded by this landscape, which for most people is a dreary, desolate, empty wasteland, Liesel once came out with this strange remark: "You know, I lived here very happily in one of my former lives. I feel at home here. If you want, I would be prepared to come and live here in a tent with you." Our visits to Jebel Egma and Jebel Et Tih were a special and remarkable experience. It was very difficult to drive up the steep slopes covered by large blocks of rock. At Jebel Egma Liesel proved that she was a master in driving a jeep. Although the army jeep, with its special gears reached the peak without any difficulty, the first command car got stuck with our jeep behind it with Liesel as the driver. Aryeh and I left our jeep to help the others, while Liesel looked for a way to drive up. She then realized that if she could pass a certain rock she would be able to go further. Going full blast, with all the wheels in the air she "flew" over the rock, ending up on the mountain. We all looked on unbelievingly. We thought that she must have hurt herself, but found that she was only a bit shaken. "That is the way to handle a jeep," she said.

During our many excursions into Sinai she became such a passionate and excellent jeep driver that El Yashiv, the commander of our army unit, a highly critical man who had originally been opposed to having a woman with us, offered to take her on as a professional jeep driver in the army. It was a pleasure to see her drive. In the sandy wadis, many of our cars got stuck, but never Liesel's. She always found the right tempo, changing gears at the critical moment in order to pass through the sandy stretches without hitting the rocks buried in the sand.

At the top of Jebel Tih and Egma, we found ourselves on the edge of a ridge which slopes steeply down to a broad valley, marking the boundary between the northern part of Sinai, composed of sedimentary rocks, and the southern mountain massif consisting of magmatic (igneous) rocks. Looking south over the valley we saw the monumental mountain massif of Jebel Mussa, Jebel Catherine and Jebel Serbal and beyond them, the glistening blue-green water of the Red Sea. The brilliant beauty of that view is truly unforgettable.

Sinai is full of traces of its long and colorful history. I came across four unknown Nabatean rock inscriptions, which were subsequently deciphered by Avraham Negev. The more interesting reads: "Be remembered Hali, the son of Halisat in welfare. In the year 161." The second one read: "Year one hundred and six which equals to the days of the three Caesars." The latter gives a rare insight into the Nabatean calendar because the year mentioned is 211 AD when Septimus Severus died, and Geta and Caracalla took over as Caesars under Severus.

Early one morning before sunrise we came across a most moving ancient monument on the high plateau of Jebel Tih. All of a sudden we found ourselves in front of two upright stone slabs looking like the stone tablets of the Ten Commandments brought down by Moses from Mount Sinai. The tablets were surrounded by a ring of stones. This site must have once been a sanctuary. Even the atheists in the group could not help feeling overawed, looking on in deep silence. Had it been built by our forefathers during their years of wandering in the desert?

On the southern coast of the Gulf of Eilat we studied the mangroves, tropical tree shrubs which grow in the sea with the tips of their roots protruding above the water. The northernmost point of their geographic distribution is Nabek, and they can be found as far south as Ras Mohammed, where they form dense stands. Interestingly, north of Ras Mohammed along the Gulf of Suez, there are no mangroves at all.

Apart from the coral reefs with their multi-colored fish, which are now well-known, there are many unique beauty spots in Sinai that the casual visitor does not usually see. At Ras Mohammed, the southernmost point of the peninsula, there is an uninterrupted view over the shimmering, shiny azure sea to the hazy contours of the mountains of Saudia and the coast of southern Egypt.

There is the strange El Bilayim lagoon surrounded by a ring of vivid green vegetation on the coast of the Gulf of Suez. With its green-blue water it looks like an oasis in a stretch of desert completely devoid of vegetation.

Last but not least there is the huge magmatic mountain massif of Southern Sinai crossed in all directions by many multi-colored dikes, intruding into the older massif, sometimes extending for miles along the ridges of the mountains, looking like the spinal cord of a great saurian. Nobody knows exactly where the Ten Commandments, the basic moral laws of mankind were given to Moses, but one feels that this event could have taken place on any of these awe-inspiring mountains.

With army permission, we once went to see the uninhabited island of Tiran, near the Gulf of Eilat, opposite Sharm El Sheikh. We went there in a flat-landing craft that I had become familiar with in the Second World War, which tends to make people feel seasick even in calm waters, as they do not have a keel. Liesel only avoided this fate by lying down on a sailor's bed which had been offered to her.

Tiran was of special interest because, as far as we knew, the botanical life in this area had never been explored. The rocks of the island consist of a mass of fossilized sea shells and snails. With my limited geological knowledge, it seemed to me that these fossils were remarkably similar to the living animals along the shore indicating that the island had been formed in relatively "recent" times.

With all the wonderful experiences we had in the Sinai there remains only one sore point. Dayan's plans based on our suggestions to use the peninsula to help to solve the Palestinian refugee problem did not materialize. Not being a politician, I can only assume that his ministerial colleagues objected to the idea, which in any view was a tragic mistake, the consequences of which we are still suffering from today.

When our investigation of the Sinai came to an advanced stage where my participation was no longer necessary, I accepted an invitation from the University of New South Wales in Sydney, Australia, in the framework of an exchange program to get to know Australian scientists working in the large desert regions there. The main aim was to develop a desert research program, which could be of mutual value. On the flight to Australia passing over Iran, I observed many kilometers of *kanats*, much longer than those I was familiar with in the Aravah.

On December 29, 1970 we arrived in Sydney, and were met by Chaikin, the Dean of the Faculty for Applied Science. We soon became good friends. We lived in University accommodations. Jack Mabbut, the well-known desert geomorphologist organized the research program. Most of the Australians with whom we came into contact, received us with open arms and a kind of hospitality which we met rarely in any other place. This was obvious in the many pubs. We, the strangers, were always treated to glasses of beer with the call of somebody shouting "it's on me," meaning that he was ready to pay for a round.

We soon found out the English spoken by Australians is somewhat different from that spoken in England and the United States. Initially we had some difficulty in understanding it. In particular the Australians pronounce the letter "a" like the German "ei," and I could understand Latin plant names which my botanical colleagues pronounced in the Australian way, only when they wrote them down.

Before starting the desert work I spent some time getting myself acquainted with the Australian plant world. I knew that the flora there is so different from that of other continents, but knowing plants from books and seeing them is not the same. On our first excursion, two botanists, Julie Burrel from Mabbut's Department, and Nola Hannon, showed us a very large Eucalyptus tree near Sidney. They told us to examine the leaves and tell them if they were all of the same shape. We could not see any difference. They then pointed out a parasite of the mistletoe family with leaves that were indistinguishable from those of his host. This strange phenomenon of convergence is not restricted to this case. It applies for example to the leaves of hundreds of different species of Eucalyptus trees which are all similar to each other in spite of their wide range of habitats and growth forms ranging from deserts to semi-tropical regions.

We came across another case of this on another trip in a place with European heath-like vegetation. Julie and Nola could not differentiate between the shrubs belonging to four different families all carrying small leaves that looked like those of heather (Erica). Since they were not in flower, neither botanists could identify them.

What is the explanation for this strange phenomenon, which I called the "similarity syndrome"? Why is it particularly widespread in Australia? I recalled Frits Went's so-called "flying genes" theory, which everybody thought was crazy at the time. However, in view of the fact that we now know that some viruses can carry genes to plants that are capable of modifying their genetic makeup, this hypothesis seems less far-fetched. Since techniques have now been developed introducing foreign genes into plant cells, maybe insects carried out a similar process.

For a botanist, Australia is not only interesting because of the "similarity syndrome". There are many more strange botanical phenomena of which I only mention two more.

During a trip to the Snowy Mountains I found, to my surprise, along a road side a considerable number of plants which I could identify without referring to my guides; I knew them well from the

Mediterranean area. They were all relative newcomers to the Australian flora. Our guides explained that over 25 percent of the plant species in today's Australia are foreign plants that originally came from the Mediterranean area, California and South Africa, and are gradually replacing the indigenous flora. Since 1836, three to four new species a year invade Australia and New Zealand in this way and this phenomenon is on the rise and endangering the local vegetation. I had come across something similar in the Argentinian pampas. In my view, none of the many theories put forward satisfactorily explain this strange phenomenon.

Our guides showed us an "Australian jungle" with Araucarias, tree ferns, epiphytes and lianas in a humid gorge in New South Wales. Although it contained different species from those in the rain forest of Missiones, the general impression was similar. Since the rainfall there is much less than in Missiones, I would never have expected to see this type of forest in Australia. Is it a relic from a much more humid period in the past?

During our stay in the wide, open spaces of the central region of the country, we stayed at the University of New South Wales Desert Research Station at Fowler's Gap. We explored the Great Victorian, Gibson, Simpson and Great Sandy deserts under the guidance of various colleagues, including Ray Perry from the CSIRO (Commonwealth Scientific and Industrial Research Organization), Peter Milthorpe, and others. In spite of many common aspects, the Australian deserts differ from those in the Middle East, North Africa and the United States. I mention only two. One aspect is the "Gilgais" of loam which swell up when wet. These swelling loams form landforms in which the loam hillocks alternate with depressions, which are devoid of vegetation. The banded vegetation pattern, which I came across near Broken Hill, of shrubs and trees growing in strips rather than more or less equally spaced, is another salient feature.

We also learned something about the strange fauna of Australia. Often, while driving at top speed over the dirt tracks, a frightened herd of emus raced alongside the car. Sometimes if we tried to get away from them, they would senselessly run into a fence, injuring themselves. They are also apparently quite curious. If we stopped the car and waved handkerchiefs at them, the emus would come up to us to see what was happening.

The kangaroos seemed to be ubiquitous. On one occasion, our guide showed us what looked like a tree trunk, which turned out to be a kangaroo. Jokingly, he pointed out that tree trunks with wriggling ears are kangaroos! The most we learned about kangaroos while staying in Fowler's Gap, was when we met a group of Australian zoologists who were studying the behaviour and social life of kangaroos, about which very little was known at that time. They found out that they live in small groups, the mother with a developing fetus in her pouch, and several other offspring by her side. The male kangaroo is rarely seen together with his spouse. The zoologists marked kangaroos by placing bands around their necks (banding), and studied their reproductive patterns. They found these to be very different from those of other mammals. The fetus is born 30 days after conception. The mother licks its own belly and on it the embryo climbs into its mother's pouch where it sucks on the mother's tits until it reaches a certain size. When in danger, the young one jumps into the pouch. However, if it falls out of the pouch, the mother runs away. According to the zoologists, the mother does not return to search for the abandoned cub. Liesel refused to believe that. But the zoologists had collected some of these abandoned animals, thereby saving their lives. One of them, a young "joey" with silvery grey fur jumped into Liesel's lap. She fondled and stroked it which it seemed to like. We took it back to our room and when left alone it ate all the plants I had collected. Their great appetite for plants is also the reason why Australian farmers kill many kangaroos and let them rot; they are too strong a competition for the sheep. Unlike most other animals the kangaroos have no fixed time of mating. Another feature of their life is that in years with little rain, which occur often in the Australian desert, they can delay the development of the fertilized eggs until the rains come.

The manager of the Fowler's Gap station was an experienced grazier (the Australian term for stock farmer). From him I learned the most about the ways and means used by which the Australian graziers of the dry region manage their cattle station. I received my first shock when he told me that a cattle station must be at least 80,000 hectares in order to be economically viable. I was therefore not astonished when later on, we went to the Tipperary station, south of Darwin, which is one-quarter the size if the State of Israel and has a herd of 20,000 head of cattle. This brought home to me what a small country Israel is! The second shock came when I found out that the Australian grazier makes very little attempt to improve the carrying capacity of his pastures. I tried to point out that at least in some of their immense holdings water-harvesting could be advantageous to them. We also went to Buffalo Creek near Darwin, where I saw a tropical mangrove jungle on a much larger scale than the one of the Gulf of Eilat.

After we had our fill of the desert, Julie Burril took us to the arable areas of Australia (Griffith, Deniliquin, Mildura and Meerbein), where many fruit trees and vines grow and wine is produced. Flying from place to place in very small planes, which were swayed by the air pockets in desert areas, often gave Liesel the feeling that we were going to crash, and she clung to me tightly.

We stayed at local CSIRO stations, and came to admire their work. This is possibly the best organization of its kind in the world. The same can be said of the National University in Canberra, which we visited a number of times.

I was in Canberra last to attend the 28th International Congress of Orientalists. I was probably the only botanist ever invited to such a meeting. I lectured about our findings in the Negev in the Section on Ancient Irrigation Systems of the Orient. We were put up in the University's students' hostel. Liesel and I were next to an illustrious guest, one of the brothers of the Emperor of Japan, who was a scientist.

A seminar on the so-called "Green Revolution", which was a new concept at the time, was part of the meeting. Again I was the only botanist among participants from the World Bank, agronomists, and social scientists. The speakers described new varieties of field crops with extremely high yields, which, in their view, had the potential of solving the world's hunger problems. The "Green Revolution" was, so we were told, a complete success, and could feed our overpopulated planet. I could not agree with their optimism, pointing out that in developing countries, such as India, the population was increasing much more rapidly than the increase in agricultural production due to the "Green Revolution". In addition, many of the new "wonder" strains needed much more water and fertilizer than conventional varieties. Farmers in poor countries had neither the money nor the know-how to cope with this problem. The so-called "Green Revolution" would perhaps make the rich richer, but it seemed to me that most of the poor farmers in developing countries would not benefit from it. All things considered, I told them that I thought that the "Green Revolution" would probably do more harm than good. The other participants jumped down my throat, a rather disagreeable discussion developed, people calling me a prophet of doom. My only answer was that the prophets of doom in the Bible had always been right. Unfortunately it has been proven that I was more accurate in my assessment than my colleagues.

In Australia, we also came into contact with the Aborigines, who had lived in the Stone Age until the advent of white settlement. Their stone implements, like the beautiful stone axe I received as a present from my colleague Carr, resembled the Neolithic tools we found in the Negev. They did not practice agriculture, but had devised a way of collecting runoff for drinking purposes. They made holes in the rocks, which filled with runoff water. Their openings were closed by stones in order to prevent evaporation, and they sometimes cut small channels leading to these primitive precursors of cisterns.

By the time we left Australia we had fallen in love with the country, its open spaces, and its population.

After our return from Australia came one of the most fruitful periods of my scientific career. Otto Lange from the Department of Botany of the University of Würzburg, one of Stocker's former students, had carried out a preliminary study of the ecophysiology of desert plants in Avdat in 1967. We had planned that at a later time we would continue this work much more intensively.

In April, 1971, Lange came to Avdat with his assistants Detlev Schulze, Ludger Kappen and Uwe Buschbohm and a special truck equipped with the latest instruments for measuring water loss (transpiration), photosynthesis and water potential (the force of the plant for absorbing water from the soil). The truck was placed at the lower end of the farm, where we had a hut for the automatic self-registering instruments. We connected the plants we wished to investigate to these instruments by long cables. The instruments worked day and night. For eight months we stayed together in Avdat without outside interference and felt like living in a scientific kibbutz.

The results of our field work, published in 53 scientific papers, the last of which appeared in 1982, were most interesting. On the occasion of my eightieth birthday, my friends presented me with a bound collection of these papers entitled: *Ökophysiologie der Wüstenpflanzen. Ergebnisse der gemeinsamen Arbeiten in Avdat 1967 und 1971.* (Ecophysiology of Desert Plants. Results of Our Joint Work at Avdat, 1967 and 1971).

The year of 1971, with its stimulating experiences in Australia, and harmonic scientific cooperation, also bestowed another gift upon us: a new daughter. A second cousin of Liesel in America, neglected by her husband, fell ill and was no longer able to take care of her son Jonathan Kalonymos and her two young daughters. The son was adopted by his grandparents, the youngest daughter died, and Jessica, the oldest girl, who was two years old at the time, was placed in a foster home. Upon discovering that a member of the Kalonymos family had fallen into strange hands, Liesel decided to adopt Jessica. I reluctantly agreed. Being 67 years old at that time I had doubts about whether we could bring up a young child. On November 9, 1971 Liesel flew to the United States, returning with Jessica three weeks later. With her energy and drive she had convinced the American authorities that it was best for the child to come and live with us in Israel. I went to collect them at the airport, and was deeply moved by the small, very unhappy-looking child. As soon as we got back home, Jessica looked around and said, "I like this house." I then knew that everything would be all right. We later adopted Jessica. With her love and affection she soon became a real daughter and a source of great joy to us. At the time of writing, she is 19 years old and serves in the army.

Chapter 14
A Visit to Russia,
and the Institute for Desert Research

In the spring of 1972, the Soviet Academy of Sciences invited Kofish and me to attend a seminar on the Productivity of Arid Ecosystems, in Leningrad. We were the first Israelis to receive such a direct invitation from this eminent body.

Since diplomatic ties between Israel and Russia had been broken after the Six-Day War, we had to obtain our visas in the Russian consulate in Vienna, where, according to the telegram from the Soviet Academy of Sciences, the visas would be waiting for us. However, when we got there, the visas were not ready. We went back to Würzburg where I did scientific work with my colleagues. Subsequently Liesel returned to Vienna to collect the visas. Since she cannot read Russian she did not realize that in the papers she received, the consulate staff had made a mistake - my name had been put under her photograph and vice versa!

On arrival in Leningrad, the passport official looked at our visas, immediately took all our papers away, and disappeared, after ordering us with a commanding gesture, to wait in a corner of his office. After the passengers of our plane had already left his office we felt very uncomfortable there, waiting all alone. After an hour, when he returned, I asked him why we had been detained. "You have clearly forged your visas," he said. Showing Liesel her photograph with my name under it, he asked, "Are you Michail?" and then he went through the same procedure with me. In vain we tried to explain that the consulate officials in Vienna had made the mistake. He didn't believe us. After another half hour of waiting I lost my patience and started shouting. I told him that we had been invited by the Academy and that one of their representatives was waiting for us outside. After long consultations between the two men our papers with the photographs exchanged, were returned. Ironically this proved to be our undoing on the return trip. We went through the whole business again, this time in reverse. We were accused of having changed the photographs round, and all our papers were confiscated again. My explanation as to what had happened on our arrival did not help. All of the other passengers were already on the plane and again we were kept waiting. Our luggage was then searched. When my Hebrew prayer book that I always carry with me was found, I was subjected to a malicious enquiry. Had we brought any other Hebrew books with us? Did we have relatives in Russia? Had we given any Hebrew books to Russian Jews? I was furious, and based on my previous experience, I started shouting, apparently the only way to fight Soviet bureaucracy. It seemed to help, since all of a sudden a higher-ranking official appeared and returned our papers without saying another word. We got to the plane minutes before take-off.

We also soon learned something about the inefficiency of the daily life. Since I speak a little Russian, we could move around quite freely in the streets of Leningrad. The simple act of getting some coffee and cake in one of the cafes went something like this: standing in line to find out what cakes were available and their cost, having decided what to order we were given a bill and then had to queue at the cashier's desk, and then, with the receipt in hand, we stood in line again to get what we had ordered. The procedure, which took over an hour, was repeated twice in our case - once for the coffee and once for the cake. I was amazed by the fact that people waited patiently, never complaining about the situation. They obviously thought that this was the normal way to handle things. I could cite many other examples of such inefficiency, but will mention only one more.

While staying in a Moscow hotel, we wanted to change a 100 DM traveller's cheque into rubles. The cashier gave us the ruble value of one hundred dollars which was much more than we should have got and when I pointed out her mistake she went pale. She said that she should not have changed German currency and that we would have to go to the Russian State Bank quite some distance away. I complained bitterly about having to go to the State Bank to settle such a minor matter, but this did not help.

At the Bank, there was a huge queue at the cashiers' desks, mainly foreign students at the University of Moscow waiting to get their stipends. They complained, but only between themselves, about the Russian bureaucracy. Realizing that we would have to wait for hours, I approached one of the supervisors, and when he refused to help, I resorted to my usual method of shouting that they should not treat an official guest of the Academy of Sciences in such a despicable way. He became confused, and advised us to go and see the Bank Manager, an idea I found ridiculous for such a trivial matter. He was extremely friendly and helpful and even asked us about Israel, gave our cheque to one of his staff, who promptly returned with the rubles. What a monumental waist of time! This happened in 1971, when only very few foreigners were allowed to visit the country, maybe things have changed by now.

The seminar itself was very interesting. Lange and Schulze also participated, and reported on the work we had done together at Avdat. There were also several other foreign guests: one from India, two Australians, three Frenchmen, three scientists from Czechoslovakia, three Germans from East Germany and one Jugoslav, and 65 Russian participants. All the visitors from non-communist countries had to pay in foreign currency and were put up in the most expensive hotels, whereas people from the communist bloc were allocated to cheap, third-class hotels. We would have preferred to have paid less and stayed in less expensive hotels with the others, but were not allowed to do so. Both in Leningrad and Moscow, we only came into contact with our colleagues from communist countries during the lectures and had no opportunity to talk to them privately.

We even went on excursions in different buses. Only once on an excursion to the Pamir mountains could we fraternize with our colleagues from Russia and the communist bloc. We got together at a restaurant in the Varsob valley, when by chance one bus arrived early and the other left late. Slightly drunk, ignoring the guide's warnings, we had a good time with our colleagues and could converse freely. Prof Meusel from East Germany gave us a gala performance, dancing on a table as a grand finale. But even here we experienced the fear of Russian Jews to contact Israelis. The "communist" party observer was accompanied by a correspondent from Tass, the Soviet news agency. From his looks and his name, he seemed to be Jewish. He interviewed the foreign members in our party. After talking to Lange, he approached me and asked where I came from. When he found out that I was from Israel, he turned round and disappeared. As a Jew, he was apparently afraid to be seen talking to an Israeli. This did not really surprise me as I had already noticed that the Russian-Jewish participants never spoke to us in the presence of their non-Jewish colleagues, only speaking to us when, by chance we were alone with them. The other Russians asked freely about Israel and its conditions. I could understand this attitude in view of the huge piles of anti-Zionist, anti-Israeli, and anti-semitic pamphlets we saw at all the airports. I took one home as a sample called: *"Zionism, Instrument of Imperialist Reaction."* It was full of blatant lies and distorted facts. A few examples:

"Ben Gurion sends people to various countries to instigate anti-semitism in order to increase emigration."

"There is no common interest between the Zionist leaders, who get fat at the expense of the Jewish working population, whom they exploit and rob."

"The workers in Israel crawl before their masters and seek the favor of their employers at the expense of their human dignity.."

"Workers on the land in the kibbutz are exploited and can, without paying, eat and live in one of the houses belonging to the kibbutz, but they have to toil in the burning sun without any financial recompensation."

"There is ample evidence showing that Zionist leaders encouraged the fascists to persecute Jews in Germany and to arrange with the fascists hideous deeds."

These are only a few "pearls" from these pamphlets. They are so insane that it is not even worth refuting them.

Falsifications of this nature were not limited to Zionism. Our official guides also misrepresented their own history - at that time Trotsky and Stalin were "non-people," who according to the official view, had not existed. In Leningrad we went to see the battle ship "Aurora," a memorial to the Bolshevik revolution. Our guide explained that a gun fired from the ship was the signal to attack the Winter Palace, the event that brought the Bolsheviks to power. With tongue in cheek, I asked the guide, "Didn't Trotsky give that fateful signal?" She pretended not to have heard my question. While on sightseeing in Dushambe, the capital of Tadjikistan, I wondered why I had never heard that name before. One of the others in the group explained that the name of the city had once been Stalinabad. The guide, however, never mentioned the city's former name. Later, I noticed a large disused pedestal for a statue, bearing clear traces of the name Stalin. I maliciously asked the guide if there had once been a statue there in honor of Stalin, and she said that she did not know, an obvious lie.

So far, I have only described "official" Russia. In contrast, our relations with the people in the street and the scientists at the conference were excellent. As Israelis they seemed to treat us with special affection possibly to show their disapproval of the official policy toward Israel. They willingly helped us, and gave Liesel small presents.

The botanical high point of the seminar was the excursion to Tadjikistan. We pestered the guides to take us to the Karakum and

Kisilkum deserts and the desert research station at Repetek, which I had known about since my student days. After all, the whole conference dealt with deserts. We were given all kinds of excuses. Repetek had no facilities for housing visitors, and the roads were impassable. Clearly for some reason they did not want us to go there.

Even so, the excursions to Tadjikistan and the Pamirs were highly rewarding experiences. On the lower slopes of the Pamir mountains, we came across a botanical sensation: large natural forests of Pistacio trees (Pistacia vera.). The guide explained that this was the homeland of the cultivated Pistachio. We ran into the forest in order to see whether these were really the same nut-bearing trees we grew in Avdat. We could not find any difference between them and our trees. The guide's explanation of the ecological conditions there brought us to understand some of the physiological properties that we had noticed in our trees. The summers there are very hot with temperatures of up to 45°C (113°F) and the winters long and cold, with temperatures going down to -40°C (-40°F). This explains why the cultivated trees can grow in hot deserts, but also require cold winter temperatures for bearing fruit, and why the pistachio is the last of all the fruit tree varieties that we grow in Avdat to sprout leaves in the spring, and also the last to lose them in the fall. Seeing the tree's natural habitat made us realize that one has to know the history of the trees in order to understand their behavior.

We also had another exciting botanical experience in the Pamirs. After my attack of angina pectoris, on doctor's instructions, I was not allowed to climb to altitudes over 1200 meters. For this reason, Liesel and I stayed behind at the botanical experimental station of Kandara, which was in the middle of a large forest, while the rest of the party went on. The female director there, who had come across my name in some of my publications, proudly showed us the library of the station, treated us to an excellent meal and showered us with gifts. After lunch, she, together with us, explored the thick forest there. I could not believe my eyes - indigenous wild fruit trees, including almonds, cherries, apples, apricots, vine and walnuts, were growing there in profusion including the Pistachio. Apparently they originated from this area.

In all the clearings grew giant herbs, including the cow parsnip (*Heracleum*), on which I wrote my first scientific paper (4). Whereas the variety I had studied in the botanical garden of the University of Frankfurt was a small plant - the one in Kandara was over two meters tall. There was also a plant with enormous white flowers like a teacup, of the bell-flower family. I was so enthused by the station with its beautiful surroundings that I would have been ready to stay and work there. This was, of course, impossible.

After our return to Israel, we went through a very difficult period. At one point I thought that we would have to give up our work on the Negev farms owing to financial difficulties. This was partly due to a mistake on my part made years ago. In 1966, Julius Simon, a relation of Kofish's from the United States had, after a visit to Avdat, offered to help us to secure our financial future. He applied on our behalf to the Hillson Foundation in New York, which was prepared to grant us one million dollars. In 1968, Max Branwen, the trustee of the Foundation came to Israel to discuss the financial details with us and the Hebrew University. We agreed to invest the money in the States and to receive the interest on an annual basis. During the discussion, the President of the University asked me if the interest of half the sum would be enough to maintain the farm, as the University urgently needed the rest of the money for the Department of Education. I agreed, because at that time the income of $500,000 was enough to keep and develop the farms. I naively thought this was only a temporary arrangement. In March 1973, we were informed by the University to close down the farms by April 1, as we had run out of funds for that year. This was correct, because with the extension of the farms, our expenses had escalated, owing to inflation in Israel. The university refused point blank to help us and we were desperate.

Our salvation came from a completely unexpected direction. For many years, Ben Gurion, who was the States' First Prime Minister, had preached the importance to Israel of development of the Negev, which accounted for 60 percent of the total land at that time. He came several times with his wife Paula to visti us in Avdat. He liked our farming methods and instructed the Settlements Authority to build more runoff farms in the Negev. Much later, I found out that the Director of the Department opposed the idea and had blocked Ben Gurion's proposal.

After Ben Gurion's resignation as prime Minister and his retirement to kibbutz Sde Boqer, a few kilometers north of Avdat, a group of his friends started a college there. They hoped that it would develop into the Cambridge of the Negev, in realization of Ben Gurion's dream. However, it started from a more humble beginning as a combined teacher's training college, high school and field school in 1972.

Field schools are typically Israeli institutions, of which there are several in various parts of the country. Their primary aim is to give visiting teams of teachers and school groups some insight into the plant and animal life, geology, and archaeology of the surrounding area. I was appointed as the provisional Director of the Sde Boqer college, which was clearly very far from being a University-level institution. In view of this, Prof. Ernst David Bergman of the Hebrew University, who had close ties with Ben Gurion, got approval from the Israeli Committee for Higher Education to establish an Institute for Desert Research within the college of Sde Boqer. The Committee accepted Bergman's proposition, and in 1973, the Israeli government agreed under the condition that the institute would be affiliated to Ben Gurion University in Beer Sheva. Prof. Amos Richmond of the Ben Gurion University was appointed Director of the Institute, which began its activities in 1973.

In desperation about our situation in Avdat, I turned to Prof. Richmond for help. Realizing that our farms could be a valuable asset to his budding institute, he immediately provided us with the means to continue our work. At the same time, he started negotiations with the Hebrew University. It was decided that the farms would become the joint property of both universities and that the Desert Research Institute would match the income we got from the Hillson Foundation. Thus Richmond saved our farms.

Under Richmond's directorship, the desert Research Institute developed into an internationally recognized center of desert research with many disciplines, including Meteorology and Climatology of Deserts, Hydrology, focusing on methods of locating water sources and their efficient utilization, and Desert Architecture, dealing with building special houses suited to desert climates, utilizing solar energy for keeping them cool in summer and warm in winter. Also under investigation were: solar energy utilization for production of algae, a rich protein source; and in Richmond's own department: desert ecosystems (I directed this unit for some time), desert runoff agriculture (our farms) and the sociology of desert peoples.

On his visits to Avdat or when we went over to Kibbutz Sde Boqer, Ben Gurion sometimes told us interesting stories about his life, some of which are worth recounting. I refer only to what Ben Gurion said and to what I, after each meeting, wrote down verbatim in my diary:

"In 1915 Ben Zvi [who later became the second president of Israel] and I were summoned by the Turkish Governor. He asked who had elected us as delegates to the 11th Zionist Congress, which took place in Vienna in 1913. Dressed in my working clothes, I answered, "The Americans." The Governor looked at me with disgust and said, "In all the United States couldn't they find somebody more suitable than you?" Later, we were both expelled from Palestine."

"I was arrested and imprisoned in Jerusalem for some time. An Arab friend, who had studied Law with me in Costa's (Constantinople) University came to visit me and asked me why I was there. I told him that I was going to be deported from Palestine. My Arab friend said, "I am sorry for you, but as an Arab, I am very glad that you are getting out of our country. While working as a laborer in Sedjera [a Jewish Settlement in Galilee], Jews were murdered by Arabs, but at the time I was convinced that those acts had been carried out by thugs. Only hearing my Arab friend, an *intellectual,* talking in this way, made me realize that the Arab problem is much more serious than I thought, and that it will be very difficult to find a solution. For this reason, in the twenties, I tried to negotiate with the Arabs. I asked Judah Magnes [who was then President of the newly established Hebrew University] to serve as an intermediary although we had very different political views. He and Antonius [a rich and respected Arab of Jerusalem] even established contact with Hadj Amin El Husseini, who was then the mufti of Jerusalem. Mussa El Alami, another well-known Arab personality was also involved. He was a remarkably decent man. We came to a sort of agreement that a Jewish State would be established on both banks of the Jordan river under the condition that it would be part of an Arab federation, which would not include North Africa and Egypt, who, according to their view, were not true Arabs."

Another story: "In 1945, when I saw that war was unavoidable and that we would need arms but had no money to buy them, I went to the Director of the Keren Kayemet [the organization for acquiring land for Jewish settlement and for afforestation projects] and asked him how much he paid for one dunam [1000 square meters] of land. "One English pound" he answered, and I then told him that I would be prepared to sell the Negev to him for that price. "Are you crazy, the Negev is not yours; how can you sell it?" After the War of Independence, the Negev was in our hands. I went to "sell" it to him for five English pounds a dunam, as a punishment for his refusal to give us money when we needed it most."

On another occasion, I told Ben Gurion that I had worked in the herbarium of Aaron Aronsohn, who, in my view, had been a scientific genius. "Having read the 100-page letter Aronsohn wrote to Judge Mack [a prominent Jewish personality in the United States] on a visit to Zikhron, I agree with you, but, I cannot forgive him for having founded NILI [the Jewish organization that carried out espionage on behalf of the British]. When the Turks found out about this, they were going to massacre all the Jews in Palestine, as they had done with the Armenians. Ironically only intervention by the Germans, allies of the Turks at that time, prevented the complete annihilation of the Jewish community." Ben Gurion said.

Ben Gurion's wife, Paula, was known for her complete lack of inhibition in saying exactly what she thought. They once came, accompanied by body guards, to Avdat during the Passover holiday. We showed them the new developments on the farm, and then went in for lunch. "I will only eat with you if you entertain all the soldiers as well." said Paula. Fortunately Liesel had prepared food in advance for the entire eight-day holiday, and so we were able to comply with this request.

Ben Gurion was also our neighbor in Jerusalem. Once we were invited to a party in honor of the Hadassah Organisation (the Women's group responsible for the hospital carrying that name). The President of Hadassah referred in her speech to Ben Gurion as "our great *old* man." Paula interrupted her immediately: "My Ben Gurion is not an old man. Professor Dushkin [a Professor in Education and much younger than Ben Gurion] over there is old," she said, pointing at him. Dushkin laughed so much that he nearly fell off his seat. The President of Hadassah then continued, "We thank Mrs. Ben Gurion for inviting us and hope to be here again next year." But, Paula butted in again. "I have just invited her, and she is already asking for another invitation." The President went on building up to the climax of her speech "Next year when we come again...."Paula interrupted again: "It would be better to send your children as new immigrants." When she promised to do so, Paula replied, "Oh yes, in a month of Sundays."

Once, Mazar and I visited Ben Gurion at Sde Boqer on University business. Liesel, who had come with us, remained alone and went to talk to Paula, who was sitting on the steps of their hut cleaning her husband's shoes. Paula asked: "Do you have children?" When Liesel said no, she went on without a moment's hesitation, "Why not?

Is it your husband's fault? If not, why don't you go to Zondek, the best gynecologist in Israel."

On another occasion, I was sitting next to Paula on a dais in the presence of various foreign ambassadors during an official ceremony. In the middle of the proceedings, Paula suddenly took off her long white gloves and threw them on the ground, nudged me, and said in a low voice, "You see what I really think about this official nonsense."

On Ben Gurion's eightieth birthday, as is customary in Israel, I said to him: "May you live to 120." Paula scolded me. "Why only 120? With the advance of modern science, man can live for at least 200 years!" Had the great statesman lived longer, perhaps Israel would be in a better state today!

As part of the Yom Kippur War, 1973 brought many crises. The war started on the holiest day in the Jewish calendar, the Day of Atonement. It caught us by surprise while we were at our prayers in the synagogue. Some time after the war Kofish died. We had observed long before his death that he coughed continuously. Naively, we often told him to go and see a physician. "Yes,yes, I will do so" was his stock answer. We did not know at the time that he was aware that he had incurable lung cancer and did not have long to live. He only told us this the last time we saw him, shortly before his death. His death came as a heavy blow for us as I had seen him as my natural successor to take over the Negev farms.

Then came my heart attack. While on the brink of death I had the strangest experiences of my life. In October 1977, the Technical University of Darmstadt conferred the honor of Ph.D. honoris causa to me. On that occasion, I gave a series of lectures on our work in the Negev, which was subsequently published as a book by the University (37). Each lecture lasted about two hours, and was often followed by long discussions lasting for two hours or more. Apparently this was too much for my heart already weakened by angina pectoris.

After the lecture series, I went to visit my cousin Raoul Jacobs in Saarbrücken, who had promised that he would take me on a visit to my hometown Metz, less than half an hour's drive away. While sightseeing in Saarbrücken I started to feel ill. My cousin took me to see his physician. When the nurse had barely started making my electrocardiogram, she ran out of the room, calling the physician. He took one look and immediately called an ambulance to take me to the hospital, where I spent eleven days in the intensive care unit. The next day Liesel was not allowed in my room, but phoned me from the room next door. As I heard her voice, I asked her to write down the strange experience I had during the night. I dictated it to her because I thought that if she didn't take it down immediately I would not be able to believe it myself later on. "The doctors were inserting a pace-maker through my arm, which I could see in the mirror over my bed. Then something extraordinary happened to me. My "ego" (whatever that may be) left me and floated freely over my own body. I was fully conscious, since I could observe the doctor's manipulations with my body, as if I was *outside* my body and not involved. This was not an illusion as all my senses were fully alert. I was enveloped by a feeling of indescribable happiness and contentment. I don't know how long I remained in that state. My "ego" then returned and re-entered my body."

I later reported this to Dr. Zwirner, the Head of the Heart Section in the hospital, and was told that many of his heart patients had similar experiences when they were on the verge of death. Indeed, he had been invited to give a lecture on this phenomenon, "Der Austritt des Ichs" ("the stepping out of the ego"), that evening.

On the second night in the intensive care unit, I contracted double pneumonia. The doctors told Liesel that I had little hope of pulling through and let her in to see me for a short time. I said: "You were here in my room all night, sitting over there in the corner. At five o'clock you disappeared." She then told me that during the night she had felt that I was in great danger. With all her power of concentration she had tried to keep me "here". She had sat in her room together with our friends Ille and Hans Petersen, who had come from Frankfurt to be with her. At five o'clock, she felt that the danger had passed and fell into a deep sleep.

After eleven days in the intensive care unit, I stayed in hospital for another two and a half months. I developed an embolus (blood clot) in my right lung, which again brought me close to death. For weeks on end, I had to lie in bed without moving. This was difficult for me, and I remembered a poem by Horace from my school days: "Aequam momento in rebus arduis servare mentem," "Maintain an unmoved poise in adversity." I kept on repeating these lines hour after hour which helped me to keep my equilibrium.

In the course of my life I had only one other experience with no rational, scientific explanation. On a visit to my brother-in-law, Gerson Stern, who lived in the small village of Kiedrich near the River Rhein when I was 17 years old, he invited the painter Schonegg, an old friend of his, who was a spiritualist and arranged a seance. They all sat round a table holding hands. I, the young scientist, thought the whole thing was ridiculous and did not join in, but observed the procedure to see how the "deception" was carried out. After a while, the table began to move up and down. Schonegg said that this meant that a "ghost" was with us. He told the "ghost" that it could speak to us - one move of the table would mean the letter "a," two would mean "b," and so on. I looked carefully but could not see anyone moving the table with their feet. Schonegg asked the "ghost" his identity. Using the "code" the ghost answered, "I am the soul of Otto." (Otto was the name of my elder brother who had fallen during the First World War). Then Schonegg asked, "Where are you? How do you feel?" and so on. In the course of the exchange with the "ghost", I thought of a question, to which nobody, including myself, knew the answer. I remembered that in Metz, my brother had a teacher called Salomon, who had become a friend of the family. We did not know what had become of him after the war and had not managed to locate him. I told Schonegg to ask the "ghost" if he could give us Salomon's address. The "ghost" gave us Salomon's address in Freiburg, with the name of the street and the number of the house. After the seance we wrote a letter to Salomon and got an answer. The address had been correct, apart from the number of the house - the "ghost" had said 24, and it was 26!

The event disturbed me because it did not fit in with my scientific "Weltanschauung" (world view). I had to find a personal solution to get me out of this dilemma. I told myself: You want to be a scientist who deals only with things that are scientifically explainable. Apparently there are phenomena which today's science can not explain. You should therefore not deal with such events, although they apparently exist and are not a figment of imagination. From there on I did not take part in any other seances. I did not know at the time that one day I would unwillingly be touched by such phenomena. I had another strange feeling after the seance: If the souls of the dead are really in a world of their own it is not right to disturb them by forcing them to re-appear. This thought brought another of Horace's poems to mind, which apparently made such a deep impression on me that I remembered it by heart: "Tu ne quaesieris, scire nefas, quem mihi, quem tibi finem di dederint, Leuconoe, nec babylonios temptaris numeros..." In free translation: "Do not ask, it is a sin to know, Leuconoe, what the Gods have planned for us. Leave them and do not go the Chaldeens [the professional soothsayers] to ask the stars."

Chapter 15
Science, the Human Condition,
and the Future of Mankind: A Biologist's View

Eighty-one years, albeit an insignificant period in the annals of Mankind, is a long time in terms of the average life span. In my "short-long" life, I have witnessed many new inventions, which have increased Man's technical potential - automobiles, airplanes, radio, television, atom bombs, hydrogen bombs and computers. Had I lived a century earlier, I would have witnessed the invention of the steam engine, mechanical weaving loom and electricity, which at that time people considered to be technical revolutions. Although these were important technical advances, their impact on every day life cannot be compared with that of today's innovations. The further back one goes in Mankind's history, the longer the interval between one invention and the next. Thus hundreds of thousands of years elapsed between the first primitive stone implements of the Paleolithic to the more refined tools of the Neolithic era.

In my 81 years, I not only experienced a technical revolution, but also a scientific one - Einstein's Theory of Relativity, Planck's Quantum Theory, Bohr's Model of the Atom, Nuclear Physics, Molecular Biology, and Genetic Engineering.

Each new discovery demands a change of our scientific "Weltanschauung" (world outlook). As compared to recent times, the period between 1600 and 1860 only saw a few important scientific discoveries - Galileo's and Kepler's Laws relating to the Movement of Planets around the Sun, and Newton's Law of Gravity.

The roots of today's scientific revolution go back to the late nineteenth century - the theory of Evolution (Darwin and Wallace), the electro-magnetic field theory (Faraday, Clark and Maxwell), and the thermo-dynamic laws (Robert Mayer and Joule).

It is easy to plot a time curve of scientific-technical-intellectual progress of mankind from the time of homo sapiens's appearance on the stage of history. It remains a straight line for hundreds of thousands of years until man fashioned his first stone implements, learnt how to make fire, invented agriculture, and domesticated animals. With the advent of the civilizations of ancient Egypt,

Mesopotamia and China, the line rises again whereas in the period between the Greek philosophers and mathematicians, and the Renaissance, it remains more or less on the same level. From then on, it begins to go up at an ever-increasing rate until, in our time, it takes an almost vertical turn upward to become an asymptotic curve.

This curve of the development of "homo intellectualis" should be compared to the similar plot for "homo ethicus". The intellect only constitutes part of our being, which is completed by the "soul" - the spiritual, emotional part of man's nature, which largely dictates his daily actions. The ethical values governing our "moral" behavior are anchored in the soul, a term used here in the non-metaphysical sense. The curve for the development of homo ethicus, starting from the same point of departure, is very different from the one for homo intellectualis.

In emotional terms, we are not very different from the Stone Age man. The ethical values which should govern moral behavior, only date back a few thousand years. First, and most important of these are the Ten Commandments: "Thou shalt not murder," translated as "Do not *kill*" (Mark 5:19); "Thou shalt not commit adultery, Thou shalt not steal, thou shalt not bear false witness against thy neighbor, Thou shalt not covert thy neighbor's house" (Exodus 20:13-14) and "Thou shalt love thy neighbor as thyself" (Leviticus 20:18). Ancient Jewish sages identified "thy neighbor" with every human being, while the New Testament goes even further "But love your enemies" (Luke 6:35).

Later came the Golden Rule of Rabbi Hillel (75 BCE - 5 AD), as the Talmud relates "When a certain heathen asked Hillel, 'I will become a proselyte on condition that you teach me the whole *Tora* [the teaching of Judaism] while I stand on one leg.' Hillel replied, *'What is hateful to you, do not do to your neighbor. That is the whole Tora'*" (38). Put in a slightly different way, the New Testament states "Whatever ye would that man should do to you, do you even so to them; for this is the law and the prophets" (Matthew 7:12 and Luke 6:31). There are also the moral teachings of Buddha (sixth century BCE), and the rules laid down by Confucius (551 BCE - 478 BCE) in the Far-East. Thus we possess the rules which should guide homo ethicus. If these rules were to be applied, killing and war, hatred and genocide would be abolished. We would live in a world of peace and justice. However, both on the individual and the collective levels, we continually disregard these ethical values. Thus, while the curve of homo ethicus remains at the same level, the plot of homo intellectualis rises from day to day and the difference between them becomes progressively greater with far-reaching biological consequences.

The curve relating to intellectual development is based on man's capacity to think rationally. This potential resides in the brain. As Wallace, who proposed the concept of evolution independently of Darwin, put it: "In the brain of the lowest savages, and, as far as we know, of the prehistoric races, we have an organ, little inferior in size and complexity to that of the highest type." (cited in 39, p. 54). On a similar note, according to the paleontologist Gould (39), "Our large brains may have originated 'for' some set of necessary skills in gathering food, socializing or whatever; but these skills do not exhaust the limits of what such a complicated machine can do." If so, then "all that we have accomplished since then is the product of cultural evolution based on a brain of unvarying capacity." Thus, with this intellectual evolution Man has stepped out of the normal evolutionary framework, taking his intellectual development into his own hands in a process that is a million-fold faster than the normal, evolutionary process and is still accelerating.

Intelligence directed towards the invention of appropriate tools allows Man to turn the products of this intellectual evolution to his material benefit, thereby changing his environment and living conditions. Most people call this "progress", oblivious of its Janus-faced connotations - on the one hand, going forward to a better, more comfortable, easier future, while on the other hand, retreating from the pristine, primeval, and primordial. The Bible describes this in terms of the expulsion of Man from "Paradise" after being seduced by the snake, the symbol of evil, to eat of the fruit of the tree of knowledge (Genesis 3: 1-24). This first "progress" went hand in hand with the ability to discern between "good" and "evil". Man could now decide to follow the ethical code or to neglect it. Although knowledge and intelligence have given rise to intellectual progress, man has not as yet collectively applied the ethical code. For this reason, all his inventions have been used in both positive and negative ways. Thus, while the first stone implements were useful tools, enhancing living conditions, the same techniques were also employed to make weapons for killing other men. Similarly, when man learnt how to melt metals, he immediately applied this new knowledge to making weapons. At the present time, innate forces of atoms can be harnessed to create electricity, but also to make the most dangerous weapons ever invented. Man not only applies his inventions for and against fellow man, but also against his environment because he does not consider the consequences of his actions. I will give only one example which as a botanist is near my heart.

About twenty percent of the earth's surface is covered by tropical forests, which Man destroys at a rate of two to three percent annually and in some cases, such as the cedar forests in the Himalaya, the destruction proceeds even more rapidly. If this process continues at this rate, most of these forests will soon no longer exist, since they are not self-regenerating. In turn, soils in such areas will be washed away, and various indigenous animal and plant species will become extinct. However, the consequences are even more far-reaching - we are destroying the so-called "lungs of the world" that contribute toward the maintenance of the temperature and carbon-dioxide equilibrium on this planet.

Thus, changing our physical and social environment has disturbed the equilibrium of the *human* ecosystem. An ecosystem can only exist as long as the living components and the environment are in mutual equilibrium. In the wake of environmental changes, only forms of life which are able to adapt to the new conditions survive. Applied to Man, this means that the human species is on the way to extinction in biological terms, if he does not immediately make radical changes in his behavior. Theoretically, using his intellect, the necessary adaptations could be carried out in a conscious and purposeful fashion. However, Man acts as if the *environment* should be able to adapt itself to his needs, and not vice-versa. If we continue along these lines, we are doomed, even without the atomic bomb. So far, neither evolution, democratic governments, totalitarian regimes, *organized* religious systems, or radical religious fundamentalists, whatever their persuasions, have succeeded in changing this situation.

The only solution lies in a radical change of attitude both with regards to social and ethical values and environment. First, Man has to give up some of his so-called technical progress, forgoing certain comforts. For example, modern agriculture in all the developed countries is geared to maximizing yields through the use of the most modern, sophisticated equipment, on the one hand, and more water for irrigation, chemical fertilizers and insecticides on the other. The use of fully automated and computerized irrigation systems with sensors in the soil which turn the water supply on and off as and when necessary, has led to huge increases in production, which should mean "progress." Then why is agriculture in developed countries including Israel in a deep crisis? Why have these modern, sophisticated techniques brought environmental deterioration in their wake? Why does agriculture in developed countries have to be subsidized with immense sums for purchasing farm machinery, fuel, fertilizers, and for providing irrigation water ? Why could these techniques not alleviate the hunger problem of developing countries? The reasons are manifold. Calculations based on modern American agriculture show that the relation between energy input and output is about 1 : 10 - ten energy units are required in order to produce one unit of energy in food. In certain branches of agriculture such as vegetables grown in greenhouses, the proportion is even higher reaching 600 : 1. In contrast, in the so-called "primitive" agriculture this relation is reversed, 1 : 50! (40).

Other dangerous disadvantages of modern methods include: over fertilization which has led to pollution of ground water in many places mainly by nitrates making it unfit for use, over pumping leading to lowering of the groundwater level, seepage of sea water into the over pumped wells as happened in the coastal plain of Israel rendering them useless.

Another bane is over production in modern agriculture which means that much of the products cannot be profitably sold so that mountains of excess wheat and butter and oceans of milk accumulate. Consequently many people leave agriculture, and as a result, the governments are forced to give even larger subsidies.

The absurd "blessings" of this over production are also reflected in the media. While on holiday in Switzerland, the Swiss newspaper *Neue Züricher Zeitung,* on August 14, 1985, carried a lead article entitled "American economy faced with record yields. Predominantly negative consequences for the people involved." On September 3, 1985, the following announcement was made over Swiss radio, "The price of bread will go up by three Rappen in the wake of this year's high wheat yield."

One might have expected that this excess production would have solved the hunger problems in developing countries. However, with the exception of limited philanthropic distribution of surplus food in emergencies, it has certainly not done so.

Most modern methods of agriculture are not applicable to the developing world. The angry reaction of a group from the semi-desert province of Turkana in Kenya, who were participating in one of our courses in runoff agriculture in Avdat, when shown the sophisticated irrigation system in Kibbutz Revivim is clearly imprinted in my mind. One of them said acidly: "Do you really think that we would be able to use such methods in our impoverished country?"

I can only see one possible solution to this problem: To change our order of priorities from maximal to optimal yields, which would take into account the relation between energy input and output, financial and social conditions and the impact on the environment. Similar reasoning applies to modern industry, which has brought the so-called affluent society to the developed countries. While the rich are showered with every conceivable comfort, we are afflicted by tragic problems, including the arms race, with unlimited capacity to kill and destroy ourselves, unemployment, the increasing gap between rich and poor, a tendency, encouraged by daily propaganda in the television and newspapers to live far above our means, increasing deterioration of the environment, and the exodus from the country to the cities, which have become unmanageably large.

Scientists have played a decisive role in creating all these problems of "progress." In 1900, Theodore Herzl, the founder of modern Zionism, and a well-known writer expressed this idea in a story entitled "Solon in Lydia" (41). Solon of Athens (ca. 638 BCE - 558 BCE) instituted there wise but strict new laws, which he inscribed on tables. The citizens didn't like the laws and persecuted him. So he decided to go into voluntary exile to Lydia's King Croesus, whose fabulous rich court had attracted many Greek intellectuals and writers like Aesop, the writer of the well-known fables. One day Croesus was confronted with the most difficult problem of his reign when a young man proposed him an invention, which would eliminate all the misery from the face of the earth. As the only reward he asked in return for the hand of the beautiful daughter of Croesus. When Solon asked what the invention was, the youngster showed a pouch containing flour, saying, "This is flour which I have made myself. I have found out how to make flour without field crops, from simple materials which occur in nature in inexhaustible quantities. I can produce as much as I want and without substantial work. In this way I am able to do in one day what one thousand farmers do in a year."

When Croesus asked Aesop and Solon for their advice, Aesop proposed to accept the invention, but Solon said: "Kill him. This young man poses the greatest threat the world has ever faced. He will free man from hunger but will rob him of the best he has, the necessity to work the soil. There is no more important act than the destruction of this beautiful youngster." Unable to reach a decision, Croesus ordered the young man to distribute his flour to the poor free-of-charge. As a result, the rich grain fields of Lydia were not worked any more, and the people became lazy, unruly, and quarrelsome as they had no outlet for the energy they had formerly expended in work. The farmers, landowners and merchants were dissatisfied with the new state of affairs, and as a result, violent uprising broke out which threatened to destroy the country. Solon rebuked the young man, "This is the result of your gift. Would it not have been better to keep your secret? Follow my advice and let the people go back to toiling in the fields and working hard. It is good for them!" With Croesus' approval, he then gave him a bowl of poison to drink, "Empty it for the sake of Mankind."

Taking a lesson from this parable, experimental science should not be banned, but scientists must take social and political responsibility for their work. Nobody will ever be able to force scientists, whose main motivation is a deeply ingrained curiosity, to explore the unknown. But they must be aware of the two sides of the coin in the discovery of every new invention: it can be applied both for the benefit and to the detriment of Mankind. The laser serves as a good example: First it was used to guide rockets to their targets, and only later as a useful tool for medical purposes.

Once doctors, receiving their degree had to take the Hippocratic Oath, first formulated by the Greek physician Hippocrates (ca 460 BCE - 357 BCE), vowing to apply their knowledge only for the good of Mankind. A similar international code binding the communities of all nations, stating that scientists have the obligation and the right to determine how their findings are applied technically, should be instituted. In extreme cases, they should even have the right to prevent the practical use of inventions or findings. Governments, political and military bodies should not have the right to overrule such decisions. Such a law could be enforced by unified action by the world's scientific community, demanding also the destruction of all the terrible weapons it has helped to create. In addition, we should fight against the increasing overpopulation of our planet. These combined actions would contribute toward curbing the steep rise in the intellectual "progress" curve. Through rational thinking we could limit the negative aspects of progress. Undoubtedly such measures would involve some loss of comforts in the western world.

However, in themselves, such strategies would not set our intellectual-technical, and moral-ethical curves on the right courses and eliminate the increasing gap between the two. A radical change in attitude involving the desire for a renaissance of the inner self is a prerequisite for achieving that end.

It is written: Love thy neighbor as thyself. This simple sentence emphasizes three words: Love, neighbor, thyself. No other word has been so misused and distorted in meaning as "love". For example, the expression "make love" or "faire l'amour" implies that love is sex. What is really meant could be expressed much more effectively by the use of the word "fucking." Such hypocritical use of language has permeated many aspects of our life. The euphemism "Mr. So and So has passed away," makes me see red. People do not "pass away," they die, but we are afraid of the truth contained in that word.

What love really is, has been eloquently described by Martin Buber (42):

> "Love does not cling to the I in such a way as to love the 'Thou' only for its 'content,' its object; but love is *between* I and Thou. The man who does not know this with his very being, does not know love. Love enjoys, and expresses, ranges in its effects.... through the whole world. In the eyes of him who takes his stand in love, and gazes out of it, men are cut free from their entanglement in bustling activity. Good people and evil, wise and foolish, beautiful and ugly, become real to him; that is, set free, they set forth in their singleness and confront him as *Thou.* In a wonderful way,.... he can be effective, helping, healing, educating, raising up, saving. Love is a responsibility for an I for a Thou. In this lies the likeness of all who love, from the smallest to the greatest, and from the blessedly protected man whose life is rounded in that of a loved being, to him who is all his life nailed to the cross of the world, and who ventures in to bring himself to the unfathomable and loves - to love all men."

The word "neighbor," means *all* other people, not only those one lives close to. What does "thyself" really mean? One must know and explore oneself. Who am I? What is my inner being? What motivates my inner actions? Only through answering those questions can one change, and give birth to a new "I", which I have to accept. Only then will I be fully conscious of myself, have the self-confidence that generates peace of mind and the power to love one's "neighbor." People who do not manage to attain this level are easily manipulated by others, particularly toward hate - the archenemy of love. Historically, the great haters have always been the bane of the world, bringing murder, war and disaster upon Mankind. However individual renaissance does not suffice, unless it goes down to the roots - the education of our children. We must effect a radical change in educational values, starting from the kindergarten. Today's education is still geared to the ideals, norms and values of our industrial, progressive, affluent, wasteful and comfortable society. All too often the "education" toward killing and war starts with toys such as models of warplanes, tanks, pistols, guns, fighter planes which are sold by the thousands in each toy shop. From the kindergarten children are imbued with the concepts of our throw-away society (Wegwerf Gesellschaft) built on acquiring and producing things, which are later discarded, a policy which is "good for business." In my days, we used slate boards for learning to write because they could be reused, as it was considered an unnecessary waste to write on paper, which we only had in the higher grades for writings we had to keep.

This example may seem trivial and naive. However, this was the first step toward our disposal society and its terrible waste of raw materials.

Apart from teaching knowledge, basic skills, I have gradually come to realize what the real aims of education should be - love, self-knowledge, self-confidence, instilling the willingness to live a simple life without too many comforts, avoidance of violence and war, tolerance, and understanding with special attention to the social and physical aspects of the environment, directed toward maintenance of the delicate equilibrium. And above all, the desire to work with will power and the mobilization of all innate forces for the practical realization of these values in everyday life is of the utmost importance. I have come across many young people who agree with me, but who are aware that their teachers with their inflexible ideas are the greatest obstacle to such re-evaluation of educational goals. They should be pensioned off, and replaced by young people, in whom I have great faith.

During my long life I have learned much about society. I have become aware of the need for radical change. Over the years, I have also changed my attitude to Science. As a young man, I believed that it could solve all mankind's problems with an infinite capacity to provide explanations for all the enigmas of Nature, and reveal the absolute truth about the structure and function of Nature. Although I lost my belief in the infallibility of Science, I never lost my love for plants, which were always much more than experimental objects to me. Over the years, my personal relations with living plants, and my admiration for their perfectly functioning systems endured and grew. I increasingly enjoy their esthetic beauty and manifold forms. My love of my plants gave me strength and peace of mind. My belief in the power of Science to find the absolute truth was shaken for the first time by Du Bois Reymond (1818-1896), an eminent 19th century physiologist, who described a scientist as a man, who after working hard to open a closed door, finds himself in a corridor with many other closed doors. Every step taken to advance the so-called frontiers of science seems to turn a simple problem into a more complex one, which in turn, raises further questions.

Photosynthesis serves as a good example. As a student, I remember the professor writing the following formula on the blackboard to describe photosynthesis:

$$6CO_2 + 6H_2O \rightarrow C_6H_{12}O_6 + 6O_2$$

This formula implied that by combining two abundant, naturally-occurring substances, plants synthesize carbohydrates and oxygen. This was so simple that we were certain to be on the verge of being able to "duplicate" this process industrially. Since then, thousands of scientists have worked on this problem, and the more we know, the more complex the process seems to be! Another pertinent example is the Bohr Model of the Atom, which seemed to provide the ultimate explanation for the structure of matter, and was particularly attractive because it mirrors the solar system. However, the innumerable new subatomic particles detected since then have made the Bohr Model completely untenable and the structure of matter remains an enigma.

Thus, in basic research we are continuously faced with new facts that do not fit into existing patterns. Someone intuitively comes up with a theory that seems to hold until new facts are discovered, and it has to be abandoned. There is a continuous interaction and contradiction between accumulated knowledge and its meaning. We are dealing with a unification-discordance-synthesis spiral. Interestingly, Hegel, the father of this triodic concept of scientific "progress," and progress of human society, illustrated his system with an example taken from plant physiology: "The seed of a plant is an initial unit of life which, when placed in its proper soil, suffers disintegration into its constituents, and yet, in virtue of its vital unity keeps these divergent elements together and reappears as a plant with its members in organic union" until the whole process is repeated [my addition].

The quest for absolute truth therefore has no end, since the dialectic spiral tends to infinity. The limits of our cognitive ability are also inherent in the nature of our brain, which apparently works along similar lines as one of the products of our brain - the computer. The quality of these man-made machines, increasing their storage and combining capacity, is enhanced from day to day. We "teach" them languages, word processing, drawing, and music, and in some respects, they are more "intelligent" than we are. Can we also improve our own "brain computers"? This is only possible within narrow limits. These limits are the material and physiological structure of the brain that emerged through the process of evolution, and therefore governs our potential to think which is bound to these given structures over which we have no control, since we cannot insert new and more sophisticated silicon chips, or change the direction of electrical currents as in man-made computers. Another possible limitation in our cognitive ability is language. Disregarding the question discussed by linguist-philosophers such as Chomsky (43), if language acquisition is an innate property of the human mind, it is clear that we can only think through language. Therefore, our thinking capacity is incarcerated within the cage of the language we use, including the symbolic "language" of mathematics. Our thinking cannot progress beyond language and be "absolute." In addition, language is an imprecise tool, which is often inadequate for communication since many words have several meanings, for example the word "fired" may mean "I fired a gun," "I fired him," "I fired a kiln," "I am fired by inspiration." In other words, the meaning of the word "fired" can only be understood within its context.

Many of my colleagues do not share my view that our cognitive ability is limited. If they did they whould show humility towards Nature, whose secrets will remain a mystery in spite of all our efforts. I have often said to myself that the Scientists should investigate nature as if everything is explorable and understandable; but at the same time, bear in mind that our cognitive capacity is limited. I always tried to pass this message on to my students.

I am not sure how far my students ever accepted this; most of them, if not all, lacked any knowledge of ontology and philosophy of nature. They may have thought that what I said was outside my professional competence.

This brings me to the problem of university "education." I belong to the rapidly disappearing race of "botanists," people who, in spite of the need to specialize in certain fields, still have a *working* knowledge of most branches of botany (physiology, anatomy, morphology, taxonomy, ecology). This is so important because only thus can we *understand* the plant as a complex *unit*. Since the whole is more than the sum of its parts, the "whole" can be comprehended only by considering all the component structures and functions in their totality and their mutual interactions. I was fortunate to study Botany under Möbius, a man with wide horizons, who was an educator as well as a scientist. In private conversations he always stressed the importance of "Allgemeinbildung" (universal culture). He advised me to go to lectures on Philosophy and Social Sciences. As a result, I became an ardent admirer of Franz Oppenheimer, father of the botanist Heinz Oppenheimer, an eminent, highly progressive sociologist. For all this I will always be grateful to Möbius and Frankfurt's University.

As a long-standing Professor Emeritus, I am convinced that our universities have failed in this respect. The word "university" is derived from the latin "universitas," meaning totality. It should be their moral responsibility to educate toward this totality. But in reality the institutions are becoming increasingly organs for transmitting professional knowledge and for training specialists. If Science and Medicine continue along these lines, universities will become mere professional training schools. There is a dilemma. Clearly, in order to survive as professionals we have to train students in very specialized fields, but if this is all we do, we are providing them with a training, and not with *education.* By losing the general overview of the field, students will know more and more about less and less. We could counteract the dangerous trend of overspecialization by introducing compulsory courses on the Philosophy of Nature and General Philosophy and History of Science. Overspecialization is dangerous because it hinders us from realizing that "Organisms are integrated entities, not collections of discrete objects" and that organisms are "integrated wholes, fundamentally not decomposable into independent, separately optimized parts" (44). In my lectures in Botany to first-year students, I used to try and give them an overall view of plant life. This course was given by one lecturer and has been substituted by "General Biology," which is given by several specialists. Naturally, each of these lecturers stresses his own specialty; as a result, the student no longer perceives the united organism which is sectioned off into a number of specialist aspects. Maybe a new type of university to give students a broad education is called for. It would be most desirable that our leading politicians, and especially our ministers should study at such universities. At the moment many of them, at least in Israel, are either lawyers or former generals.

I return now to the theory of evolution, a central point of modern biology because it is an example how a matter of seemingly pure scientific concern can have disastrous effects when applied to human society. Without going into details, the theory of evolution rests on three fundamental principles. First, the world, including all living beings is not static but undergoes continuous change. Second, with the exception of identical twins and clones of plants, organisms are never genetically identical, which is the underlying reason for the infinite variety of structure and physiology of Man, animal and plant. Each human being is unique, thus not fully identical to any other human being who has ever existed, providing the biological justification for the Biblical "Do not kill." By destroying a unique human being, who might have produced another unique being, we may be eliminating a potential genius or a saviour of mankind. Third, natural selection through the struggle for existence, leading to the survival of the fittest, guarantees that the biologically most well adapted species survive.

Within the realm of Biology, the Theory of Evolution represents great progress, even if one disagrees with details. However, its immediate application to human society proved to be very dangerous. Human beings were selected and killed on the pretext that their ideas, creed, race or disabilities made them unfit for a particular society. The Nazi doctors who killed the inmates of lunatic asylums, and selected people for the gas chambers, and the infamous Mengele, the "doctor" of Auschwitz, who chose suitable people for his inhuman experiments, justified these acts on the basis of the "laws" of natural selection. *Man* could decide who was fit to live and who was not! The struggle for survival has become another dangerous instrument in human hands, used to warrant exploitation, suppression, and injustice, since the strongest have the upper hand. Who are the strongest? Men, or groups with financial resources, possessions, influence with the authorities and worst of all, pushy egoists who do not care for other people's welfare. Such people justify their actions by claiming that they are "strong" and fit to survive. For them "weak" is an equivalent of "unfit." This contention ignores the fact that the so-called weaker ones are often morally better human beings, more valuable for society than the "strong" ones.

Since an autobiography *should* be a self-confession of sorts, and not just a means for self-justification, I conclude with my deepest "non-scientific" beliefs. I was always proud of being Jewish because Judaism contains the basic values of mankind and consequently I have always kept certain Jewish ritual customs. As a young man I was an agnostic, but gradually became convinced that Man cannot live without "religion," a word derived from the Latin "religere," which means "binding" - tying the soul to "something" far out-of-reach and unfathomable, but nevertheless real. Some people call this "something" God, another frequently misused word. I do not believe in the God of organized religions - Judaism, Christianity, and Islam. Ironically *Science* has led me to believe in a force which cannot be defined scientifically. Through learning more about Science and its complexity, I came to realize that some force must bring the innumerable parts of the whole together, although it cannot be defined and is only recognizable by its effects in the same way as a physicist recognizes the existence of a force through its actions. For want of a better term this infinite force which keeps the universe going I call God in the mystical Biblical sense of "I am that I am" (Exodus 3:14). I feel He exists, I experience this force in my feelings, He touches me and I communicate with Him, but I cannot understand or fathom Him, just as Science, with its ever-increasing knowledge cannot explain the workings of the Universe. Some people may think that this belief in the absolute "It" or God, which has nothing in common with the type of "God" worshipped in the formalized religions means that I am a non-scientific mystic. This is not so since Science and my belief in His existence represent two different levels of consciousness.

I end this book with a prayer-poem, attributed to Ibn Gabirol (ca. 1020-1070 AD), the great Hebrew poet, who lived in Spain during the zenith of Arab-Jewish cultural cooperation:

> "Master of the Universe who ruled before anything was created.
> At the time when all came into being,
> His name was proclaimed King.
> And when all will cease to be,
> He alone, awesome will remain.
> And he was, and He is,
> And He will be in glorious eternity.
> And He is one and there is no second,
> To share with him, to place beside Him.
> He is without beginning, without end.
> And he is the power and the dominion.
> And He is my God and my living redeemer,
> In my travail at the time of distress.
> And He is my banner and my stronghold,
> The cup of life whenever I call.
> I place my soul within his palm,
> When I sleep and when I wake.
> And with my spirit and my body
> He is with me,
> I rest in Him in fearless calm."

Bibliography

1)   Van der Post, L., 1965. Journey into Russia. Penguin Book, 352 pp.

2)   Francé, R.H., 1912. Die Welt der Pflanze. Eine volkstümliche Botanik. Berlin-Wien, Ullstein, 455 pp.

3)   Wagner, A., 1912. Die Lebensgeheimnisse der Pflanze. Leipzig, Thomas 190 pp.

4)   Schwarz, W., 1926. Die Wellung der Gefässbündel bei Heracleum. Planta 2:19-26.

5)   Möbius, M., 1927. Die Farbstoffe der Pflanzen.Linsbauer's Handbuch der Pflanzenanatomie, 1. Abteilung. Berlin, Gebr. Bornträger, 200 pp.

6)   Möbius, M., 1937. Geschichte der Botanik von den ersten Anfängen bis zur Gegenwart. Jena, Fischer, 458 pp.

7)   Schwarz, W., 1928. Das Problem der mitogenetischen Strahlen. Biologisches Zentralblatt 48: 302-308.

8)   Schwarz, W., 1928 Zur Ätiologie der geaderten Panaschierung. Planta: 660-680.

9)   Schwarz, W., 1927. Die Entwicklung des Blattes bei Plectranthus fruticosus und Ligustrum vulgare und die Theorie der Periklinalchimaeren. Planta 3: 499-526.

10) Kropotkin, V.A., 1920. Gegenseitige Hilfe in der Tier- und Menschenwelt. Leipzig, T. Thomas, 318 pp.

11) Becher, E., 1917. Die fremddienliche Zweckmässigkeit der Pflanzengallen und die Hypothese eines überindividuellen Seelischen. Veit u. Comp., Leipzig, 149 pp.

12) Schwarz, W., 1929. Zur physiologischen Anatomie der Fruchtstiele schwerer Früchte. Planta 8: 185-251.

13)   Schwarz, W. 1929. Der Einfluss der Zug-, Knick- und Biegungsbeanspruchung
      auf das mechanische Gewebesystem der Pflanzen. Beihefte zum botanischen
      Centralblatt 46: 306-338.

14)   Oppenheimer, H., 1930. Reliquiae Aaronsohnianae I. Florula Transjordanica.
      Bulletin de la Société Botanique de Génève 22: 126-409.

15)   Oppenheimer, H. and Evenari, M., 1940. Reliquae Aaronsohianae II. Florula
      Cisjordanica. Bulletin de la Société Botanique de Génève 31: 1-431.

16)   Hitler, A., 1934. Mein Kampf. Franz Eher Nachf., München, 102-106.

17)   Feder, G., 1919. Das Manifest zur Brechung der Zinsknechtschaft des Geldes.
      Franz Eher Nachf., München, 62

18)   Hitler, A., 1939. Mein Kampf, complete and unabridged English edition.
      Reynal and Hitchcock, New York 1939, 991 pp.

19)   Schwarz, W., 1933. Die Strukturänderungen sprossloser Blattstecklinge und ihre
      Ursachen. Jahrbücher für wissenschaftliche Botanik 78: 92-155.

20)   Schwarz, W.,1931. Beiträge zur Entwicklungsgeschichte der Panaschierung. I:
      Entwichlungsgeschichte der Plastiden einiger grüner Pflanzen. Zeitschr. f. Bot.
      25: 1-57.

21)   Maximov, N.A., 1929. The Plant in Relation to Water. A study of the
      physiological basis of drought resistance. London, George Allen and Unwin
      Ltd., 451 pp (first published in Russia 1926).

22)   Volkens, G., 1912. Die Flora der aegyptisch-arabischen Wüste auf Grundlage
      anatomisch-physiologischer Forschung. Berlin, Gebr. Borntränger, 156 pp.

23)   Stocker, O., 1928. Der Wasserhaushalt aegyptischer Wüsten- und Salzpflanzen
      vom Standpunkt einer experimentellen und vergleichenden Pflanzengeographie.
      Bot. Abh. 13: 1-200.

24)   Evenari, M. (Schwarz, W.) and Richter, R., 1937. Physiological-Ecological
      investigations in the Wilderness of Judea. Journ. Linn. Soc. Bot. 51: 333-381.

25)   Molisch, H., 1930. Pflanzenphysiologie als Theorie der Gärtnerei. 6. Aufl.,
      Jena, Gustav Fischer, 337 pp.

26) Evenari, M, 1940. Germination inhibitors. Bot. Review 15: 153-194.

27) Lang, A., 1980. Some recollections and reflections. Ann. Review of Plant Physiol. 31: 1-28.

28) Galbraith, J.K., 1955. The affluent society. Boston, Mifflin Co., 368 pp.

29) Galbraith, J.K., 1956. American Capitalism. The concept of countervailing power. Boston, Mifflin CO., 208 pp.

30) Galbraith, J.K., 1955. Economics and the art of controversy. New Brunswick, Rutgers University Press, 111 pp.

31) Kollek, T. and Kollek. A., 1985. Ein Leben für Jerusalem. Wilhelm Heyne Verlag, München, 431 pp.

32) Dubnow, S., 1926. Weltgeschichte des Jüdischen Volkes, Vol. IV, Jüdischer Verlag, Berlin, 137 pp.

33) Palmer, E. H., 1876. Der Schauplatz der vierzigjährigen Wüstenwanderung Israels. Friedr. Andreas Perthes, Botha, 460 pp.

34) Diodorus Siculus, 1962. The library of history. Vol. X, books 19 and 20. Loeb classical library, Harvard University Press, Cambridge. (Original Greek text and English translation).

35) Kraemer, C.J., 1958. Excavations at Nessana, Vol. 3. Non literary papyri (Colt) Archaeological Institute. Princeton University Press, Princeton.

36) Evenari, M., Shanan, L., Tadmor, N., 1982. The Negev. The Challenge of the Desert. Harvard University Press, Cambridge and London, 437 pp. (second enlarged edition of the book; first edition published in 1971).

37) Evenari, M., 1982. Ökologisch-landwirtschaftliche Forschungen im Negev. Technische Hochschule Darmstadt, 219 pp.

38) The Babylonian Talmud. Seder Moed, Chapter Shabbath. 31a (p. 140), Soncino Press, London 1938.

39) Gould, S.J., 1980. The Panda's thumb. W.W. Norton and Company, New-York - London.

40)   Kress, K., Mikelskis, H., Müller-Arnke, H. and Reichenbacher, W., 1984.
      Energie. Verlag Diesterweg und Sauerländer, 219 pp.

41)   Herzl, Th., 1900. Philosophische Erzählungen. Verlag Gebr. Pätel, Berlin, 244
      pp.

42)   Buber, M.,1937. I and Thou (translated by R.G. Smith) T. and T. Clark,
      Edinborough, 120 pp.

43)   Chomsky, N., 1967. Recent contributions to the theory of innate ideas.
      Synthese 17: 2-11.

44)   Gould, S.J.. and Levantine, R.C., 1979. The sprandrel of San Marco and the
      Panglossian paradigm, a critique of the adaptationists programme. Proceedings
      of the Royal Society London, Ser. B. 205: 581-598.